CONTE
EDINBURGH REVI

TWICE YEARLY: AUTU

more boreal

Editorial	3

Stories of a Kind:

Margaret Elphinstone: *Stone Circles*	11
Mikhail Prishvin: from *Dripping Forest*	17
Angus Dunn: *Andrew's Teeth & the Nature of Reality*	21
Jørgen Frantz Jacobsen: *Travail World, Farewell!*	27
Gregor Addison: *Gaelic Lessons with Elgar*	40

Branches off the Language Tree: circumpolarpoetry
siusaidh nicneill: Rody Gorman: Ian Stephen: Kevin MacNeil: Anne MacLeod; Ian McDonough: George Gunn; Harry Gilonis; Alastair Peebles; Alex Cluness: Roi Pattursson; Jean Morriset; Andrew Suknaski: Susan Musgrave; Gennady Aygi: Yunna Morits: Mikhail Prishvin; Nils-Aslak Valkeapaa: Inger & Paulus Utsi: Kirsti Paltto: Rauni Magga Lukkari; Lauri Otonkoski: Markus Jääskeläinen; Pär Lagerkvist: Tomas Tranströmer: Eva Ström; Tarjei Vesaas: Sigmund Mjelve; Christine de Luca; Knut Ødegard: Andrew Greig; Jeffrey Shotts

More Boreali:

Ian Stephen & Murdo Macdonald: *Vital Memorials*	115
Roni Horn: *I Go to Iceland*	120
Gavin Renwick: *More Borealis: a Canadian travelogue*	124
Wa-Sha-Quon-Asin & Anahareo: Grey Owl's Trail	131
Barry Lopez & Alec Finlay: an interview: *Agape*	140
Allice Legat: *Travelling Back Through Stories*	147
Emily Carr: '... while the Missionaries prayed'	156
Ian McKeever: *Gradually Going Magadan*	161
Tony McManus: *Dwelling in the North*	165
Reviews & Shortleet	173

**Edinburgh Review
22 George Square, Edinburgh EH8 9LF
tel 0131-650 6206
fax 0131-662 0053**

editors	Robert Alan Jamieson
	Gavin Wallace
advisory editors	Jackie Jones, Murdo Macdonald, Alec Finlay
cover	Neil Christie
production	Pam O'Connor
publicity	Judith Scott
logo	Alasdair Gray
©	the contributors 1997

ISSN 0267 6672   ISBN 0 7486 1138 X

distributed in the UK by Edinburgh University Press
typeset in Sabon by Koinonia Ltd, Bury
printed and bound in Great Britain by
by Page Bros Limited, Norwich

subsidised by the   THE SCOTTISH ARTS COUNCIL

back issues available
from Edinburgh University Press
22 George Square,
Edinburgh EH8 9LF
tel 0131-650 6206

EDITORIAL

MORE BOREALI:
(*Latin: in a northern way*)

> I'd like the Scottish Nationalists more if they didn't have the word Nationalist in their title, I said. Or Scottish. That's only there to give them an excuse to slag off the English. I thought for a second. I reckon they should call it the Land and Water and Trees and Rain and Buildings Party. That way nobody gets left out.
>
> Pat Hunter, in Duncan McLean's *Blackden*

WE ARE OFTEN TOLD this is a small country, out on the Europeriphery, looking habitually south and slightly eastward to the golden triangle connecting London, Paris and Bonn/Berlin, with Brussels at its midst – once the high road to England was the finest sight to be seen; now maybe it's the Channel Tunnel.

And it's said that this is not only a small country, but that it's mostly a wild, uninhabited one. Since the industrial revolution, the people have cooried ever closer to the great rivers, abandoned or been cleared from the higher ground. Between the Border hills and the Grampians, more and more of the good arable land is being built on, more sprawling roads spread, more branches of McDonalds and multi-national clone-factories open, while the bread and butter industries have died and one-work communities, hit by closure, shut up shop. In the far north and west, whole villages have disappeared, whole islands have been deserted and the wilderness has reclaimed them.

And we are told that ours is a sick, repressed society, fatally flawed, split off or hanging on by the fingernails to a tilting sanity; either a Bible-ridden bunch of forelocker tuggers, a dislocated population housed on their own reservations, boozed up and furniture screwed down in case a junkie tries to nick it, or suburbanite dog-professionals attending Gaelic wailing and spinning classes, undergoing nervous breakdowns to recover some mythic race-memory. A sick country: high cholesterol, bad liver and a withered hand. The only happy ones, it would seem, are those content to be Tories, because at least they're rich even if they are fucked up by their parents, and can afford to blast some innocent creature, or fly off to OZ when SAD strikes.

Yes, the story goes that this is a country of seasonal swings, of tribally divided selves, Tcheuchter an Sassenach, Tim an Billy, splits an schisms – where one hand does what it likes and the other is too timid to do anything about it – Dr Jekyll, antisyzygies an a' that. Too variegated, too set against ourselves, for any simple nation-state identity to fit neatly.

The young Aberdeenshire loon Pat Hunter's idea of reducing the SNP to a 'Land and Water and Trees and Rain and Buildings Party' in Duncan McLean's novel has an appeal as a way of ducking out of traditional tribalism, of avoiding exclusivity. And a 'Land and ... party', one that puts the interests of the natural world at least uppiesides with the human, does perhaps exist in some nebulous form, a number of single issue groups, yet to accrete around Greenpeace, the Green Party or Friends of the Earth. In the new Scottish Parliament, Green representation via PR is a distinct possibility.

Right now, we suspect that a 'Land Party', as ownership stands, would most likely tend towards the days before the 1832 reforms, and that absentee landlords would take their seats once or twice a year to pick up their expenses. But in one thing the loon Hunter got it dead richt – calling a party a name like Scottish Nationalist limits; it recalls the party's origins in the spurious pre-World War II debates (involving writers such as Buchan, Gunn, MacDiarmid and Linklater) as to whether the Norse or Celtic element in the Scottish race was dominant, recounted in the recent *Scottish Skalds and Sagamen* by Julian D'Arcy.

Out of these debates, the need for a unifying political banner became apparent to some, and Scottishness provided it – not to keep other people out, so much as to keep the people of Scotland in. To hold the haill clamjamfrie, there was manufactured some kind of one-size-fits-all seamless garment of communal Synthetic Scots identity. In this disguise, the gallus Glesca Bear and the green-Hibernia Leither are taken to be the same man deep down, and if you peel him further, underneath you'll also find Donald o the Isle of Skye seeking his troosers, and Johnnie Gibb o' Gushetneuk coontin his penga. All bound together by something about the light, the prevailing winds, songs learned at mammy's knee, a lilt in the lineage even if the Gaelic's lost; an almost instinctive wariness of the more sophisticated and powerful neighbour to the south; and the knowledge of the safe place of retreat to the north: iconic but'n'ben i' the glen, or mouldering castle in silence-inspiring emptiness. Just as there is a myth of the evacuated north, so the mythology of the north has been evacuated – where it rises in recent memory, rapine in the Third Reich's iconography, it appears distorted and ugly beyond countenance. The wee folk of the north still resent that misappropriation.

Yet still and all, a London prince with a Scottish grannie and a Fascist grand-uncle may develop a noble friendship with a humble Lewis crofter despite their differences, and some deep awareness of the land binds all, somehow in the blood. Or so the story goes, as if southern urban Scotland/Britain had been made out of the raw material of the north, had absorbed it and all it had to offer, as if the north had been left behind, yet somehow remained a constant in the soul, impressed like the remembered image of a favourite mountain ridge. The myth of the empty north, synthesised in the southern soul, is now widespread.

Yet there are plenty of folk living in the north of Scotland who know well enough that simple things like mountains make a difference to quality of material life, and are not only there for the benefit of the soul. Travelling around them, many miles there and many miles back, to buy a pair of shoes or replace a broken tool is not the normal experience of a central-belt dweller. On the north and western peripheries, the usual nearby amenities are not so usual or nearby. For many, life is lived at a distance – very often away from other people altogether. Children are collected up at a ripe age and taken to be educated at the nearest large town, some for a term at a time. Folk travel long distances to school and work in the course of a week – in fact, the way of life begins to sound like that of a Norfolk-to-London commuter, whose children are off to boarding school. But these peripheral commuters aren't earning 100K pa. – unless they own Norfrost. Recent research into poverty in Scotland has shown that there are folk in the Highlands and Islands living below the poverty line in sub-standard housing, as poor as any conditions in the lamented 'reservations' around the central-belt cities, struggling against the highest cost of living in the country, with a barely-existent public transport system. Indeed, the similarities between the geographically northern 'rural peasant' experience described and that of 'urban peasants' living 'peripheral' lives around the southern cities are not to be overlooked. Despite obvious differences, they have 'Common Cause' across the tribal divide.

*

> Something really does happen to most people who go into the north. They become at the very least aware of the creative opportunity which the physical fact of the country represents and come to measure their own work and life against that rather staggering creative possibility.
>
> Glenn Gould

It seems, at this point in its history, as if Scottish culture must 'go into the north', to fulfill the creative potential of its future as a northern European country in post-imperial Britain. For when the centre of government shifts, its relationship with the 'peripheries' changes – an Edinburgh parliament will obviously alter the political map by shifting power northwards.

In the extreme case of 'Independence in Europe', Scotland stands to Europe, equivalently, as Lewis does to Britain: a small, sublime place out on the north-west, speaking strange linguistic forms, possessing a fascinating ancient history, having provided a good many inventive and skilled for the long gone Empire – as well as proportionately high numbers of dead in the great 20th century wars. A good place to site a military base, a nuclear dump or an oil terminal, maybe the least-damage option for an environmental disaster?

In such a changed Euro-situation, the 'peripheral' northern experience of boom/bust capitalist exploitation at the hands of external economic forces is more relevant now to southern Scotland than during the years of the Ravenscraig syndrome. Alliance with 'peripheral' neighbours around the rim of Europe can provide a safeguard against the sublime isolation of a 'euro-Lewis'. In so doing Scotland realises its position as a European country, rather than a British region, rejoining the lines of orbit around the golden triangle at a valency akin to that of the Scandinavian countries, of Eire, of Portugal, Catalonia, southern Italy, Greece, Poland, Russia and so on – creating the rim of the Euro-wheel, where Europe touches the rest of the world.

Edinburgh Review 98 'goes into the north': fixes its focus on the geographically northern frontier and, by placing Scotland in that context, a shared existence lived across national and tribal borders – an ancient internationalism, dating from a time when modern political centres were small or non-existent. We wish to engage the reader with the Canadian composer Glenn Gould's idea of the north and to extend that as a metaphor suited to describing our view of the post-devolutionary mind-set required to make the new Scotland. We wish to emphasise that 'staggering creative possibility', to free 'north' from literality, so that as a pure symbol it might mean those areas in which creativity is most challenged by possibility, therefore to include urban deprivation, not merely virgin territory. We wish it to stand for those areas of northern wilderness *and* the wasteland of post-industrial poisoning scarring large areas of the country and its seas. In both instances, land ownership is the key to use or abuse. We agree with Patrick Geddes that 'each must cultivate their own garden'. But this presumes the possession of a garden.

Now, more than ever, Scotland's people must 'measure their own life and work against that rather staggering creative possibility' – to paraphrase Alasdair Gray, work *as if* in the first days of a better situation and make it so, not because of place of origin, but a common commitment to fulfilling the creative potential of their locality. As the Liverpool Daily Echo said in 1938 of trapper-turned-conservationist-and-Indian spokesman Grey Owl when he was unmasked, as not a 'Scotch-Apache halfbreed' as he claimed, but none other than plain Archie Belaney from Hastings, Sussex, 'What after all does ancestry matter?'

> ... When a man has devoted his best years to such a cause it is surely unfair that he should be dubbed an impostor because he may ... have been an Englishman and not a Red Indian ... the record of his work speaks for itself.

Scottishness, that variegated, synthetic thing *is* desirable – tourists carry away enough tartanrie in suitcases. Moves towards clearer definition of Scotland as a political entity can only increase that interest. And for those cultural refugees, Scots-born or not, who want a total covering – to be part of Scotland – it should be acquired slowly, like a long plaid of egalitarian ideas swung around a shoulder, containing threads of all the colours, fitting regardless of shape or hue. Belaney's story is a lesson to those who would wear that plaid of many colours – only sincere concern for the future of his chosen home saved the exposed 'Grey Owl' from the midden of great historical pretenders.

In post-devolutionary Scotland, the differences which exist between conditions and cultures in areas of the country must be recognised, by further devolution of power if desired. An Edinburgh government prepared to ask the people of a place what help they require to continue to survive, and hopefully to thrive, rather than doling out subsistence from central office or encouraging evacuation, is one which recognises that these 'northern' lands are not empty at all, but full of creative potential when examined closely. In any territory, small or large, it is by developing a sense of 'integrity', of being 'a centre', by ridding itself of that term 'periphery' – often merely a euphemism for 'overlooked'/'unimportant'/'impoverished' – that the devolutionary process unfolds.

Let the 'north' of Scotland remain a place where tradition and history are valued, not swept away by developers – but not as a heritage park for tourists, where the name McDonald immediately conjures up a Happy Meal. It isnae Disneyland – real people live here too. And let those real people – whether they speak Gaelic, Scots or English, or are among the 95,000 who have none of these as their first language – let them live in real housing suited to the environment, connected to the world beyond without penalising taxes, tolls or unsuitable laws enforced from outwith, free from the laird's whim-dictation, in a north where children can be educated close to home, sure in their local wisdom as well as trained to live in the newly unifying global continent of Technowondaland. For as the Dogrib elders of the Canadian North West Territories say, in Allice Legat's essay published here, such 'strategy' makes the young 'strong like two people'.

As Scotland further separates itself from 'the Yookay', the irony that the spread of so patently non-English writing laid out in this issue of *Edinburgh Review* should be largely in forms of the English language is not lost on us. As with the language of earlier great empires, even peoples who have never fallen under its dominion seek a mastery of the imperial tongue, and desire to hear their local discourse translated into that mainstream because, as with Latin, the availability of such a *lingua franca* enables an increased exchange of learning. Ironic, perhaps, but an important attunement and a continual reminder of the historical conditions which have created our current situation. For the Dogrib Dene, to use the imperial tongue to win back their land is surely justified as an act of cultural vengeance; while to do so by utilising the wisdom of a separate tradition is indeed to be 'strong like two people'.

Let Scotland's future be a 'north' in which 'strong like two people' links with its boreal neighbours are positively encouraged, so that we may learn from each other. Our links with the south are strong already – too much so, judging by the referendum, as if the handclasp has become a stranglehold. By looking north, that habitual south-eastern crick in the Scottish neck clicks back into place. These are not idle poet's dreams – not even ideals. They are the policies and practices of a northern European country entering the new millennium, governed *more boreali*.

More Boreal is the last issue of *Edinburgh Review* to be co-edited by Robert Alan Jamieson and Gavin Wallace. The outgoing editors would like to express their gratitude to Pam O'Connor at Edinburgh University Press for her patience and support during a time of change, and to Alec Finlay for his hard work and willingness to step purposefully into the resultant void.

At the time of going to press, plans are laid to situate *ER* within the new Centre for the History of Ideas in Scotland at the University of Edinburgh – in Buccleugh Place, a stone's throw from its original source. The editors take great satisfaction in having brought *ER* 'home' again, after a period of protracted, but productive wandering. Departing now, in a devolving Scotland, their hope is that *ER* will continue to place Scotland profoundly 'in the world', as it has done through many previous editorial changes.

STORIES OF A KIND

MARGARET ELPHINSTONE

MIKHAIL PRISHVIN

ANGUS DUNN

JØRGEN-FRANTZ JACOBSEN

GREGOR ADDISON

Stone Circles

Margaret Elphinstone

NO, I'VE NO story to tell. You say it's the last chance, but nothing final is about to happen, and there's no such thing as chance, and so your reasons are so much air to me. Why do you want words from me? Do you know what words are? Nothing: a breath, a sound, a volatile spirit. I've made what is certain and eternal, which is more than most men are able to do. I'm satisfied. If you want to find a meaning somewhere, you know where to look.

Very well, I'll give you till supper time. After that, if it's not too much to ask, I would like some time by myself.

Because there's no point in talking to you. I mean that at its face value, not as an insult. We don't communicate, you and I. I communicate far better by myself. The best exchanges of all are those one makes with oneself; there is so little space then for error. That's why I want time tonight for one last dialogue. I shall miss solitary conversation. I shall exist, you understand, after the sunrise is over, but I shall be one.

No, a man is not two. He is at least three, and probably far more, only the further selves are beyond apprehension. I've thought for a long time that the differences between one aspect and another approximate to the number of the moving stars. I have no proof. I've been limited for too long by the impressions made upon my five senses, so what can I say about the things that lie beyond their comprehension? Words are made by breath, which is often foul, and by sound, which is deceptive. Why don't you leave me alone and study what is eternal? I have put a mark upon the earth which will be here until the end of the world. Isn't that enough for you? If you want words, young man, go and find a woman, or a boy, if you like that better, and get them to say to you, 'I love you'. Wouldn't that suit you better than disturbing a man who will only be what he is until the morning?

The end of the world will occur in four thousand years from now. Does that make you feel better? No, you can't imagine four thousand years, and nor can I. Don't try. I've lived for forty years, which is a good lifetime, and you for scarcely half that. It's not in our bodies to know four thousand years. How did I arrive at that figure? I calculated it from the stars.

Of course, the stars. There's nothing else to read. Nothing on earth lasts for long enough, and there are no points in the sea, nothing specific. You can't read flux. Only the stars make patterns which are certain and eternal, and men can't reach the stars. They try. That's why they travel, that's why they always seek out something that they haven't got. Because it's really the stars they seek they will never, of course, be content, since there's no boat that can carry a man to the stars. They are in every possible way beyond us.

Are the stars gods? What a foolish question. There are gods everywhere. Gods in trees and rivers mostly, who live for a short time while people give them names. There are Sun and Moon. Yes, have them for gods if you like, if it makes you happy. We offer them as gods to the common people. I'm not one of those, and I see you too have aspirations.

Very well, but if you're not ready to hear what I tell you I warn you it will make you mad.

It's your responsibility, not mine. Look at it this way. There are gods everywhere, in the stars and the sun and moon if you wish, but they're not important. Even stars can die. I've seen it happen. There's only one thing which is there for ever, which is eternal because it changes all the time. That is the pattern which the stars make. There are no patterns on earth, but there are men. We can make patterns, some of us. The common people say magic, but patterns are what magic is.

People come to see what I've made, and they look at the stones. But stones may fall. Lightning can crack a stone in two. The pebbles on the shore were once great stones, but the sea tore them apart. No, stones aren't eternal. But they are made of something that lasts for ever, and out of such material a permanent pattern can be made. A pattern on earth, you understand? It's the only thing that can make this life endurable.

My life? Why ever should you want to know about that? Don't you see how completely irrelevant it is?

Oh, I begin to understand. Perhaps you're right, perhaps there's a pattern in it. Maybe you're becoming what I've been, and maybe that's a good enough reason for telling you.

I was a boy as you were. It doesn't matter where, but it was on the coast, far south of this island. Our village did well because men used to come from inland with tin and furs, and others came by sea with fish and salt and copper, and so a market grew up, and a smithy where they made bronze for weapons and ornaments. I was a chieftain's son, one of six, a dangerous thing to be, especially as I was the youngest. I wouldn't have outlived my father for long if I'd stayed, but when I was ten a magician came to our village. Oh yes, we had an ordinary seer, the sort who reaches an inferior kind of trance through the beating of drums and the chanting of women, and sees as far as next year's harvest and the warband over the hill. Do you know the difference between a seer and a magician?

No, well, nor did I, at first. But the magician attracted me. I asked him to teach me the first rules of magic, and he took my hands in his and made me stretch out my fingers. 'What do you read here?' he asked me. I told him the name for each finger, and my thumb, on each hand. 'Those are words:' he said. 'One, two, three, four, five.' He laid his hand over my small hands, hiding them. 'Where are the words now?'

'In your hand.'

'No, in your thoughts. Shut your eyes. Do you see your hands?'

'Yes.'

'Good. Now there are five pairs of hands. Do you see them?'

'Yes.'

'Good. See them well. I shall tell you the name of every finger as we come to it.'

There were a hundred names, but even so it was not long before I could remember them, because each pair of hands had the name of their owner, and the finger patterns were contained in that. Only the first ten fingers needed no name, because they were mine, inside myself.

When I was master of the hundred names, the magician made me shut my eyes, and then he took the hands away so I couldn't see them any more. After that there were only numbers, as many as the stars in the sky. Like the stars, they came in patterns. I tried to count to the end of them all as I lay in bed at night, but it was like counting the stars, and whenever I came near the horizon the circle of the sky rolled onward, so there were always more. I could not reach the end, and so I gave up, and concentrated on the patterns instead. My master taught me all he could during the seven years I was his pupil, and I learned that there was no end to the patterns it was possible to make.

On clear nights we studied the sky, because you have to look out of the world to find a mirror for your thoughts, just as you look into water to find a mirror for your senses. But the real magic was dark. Just as he had done on the first day, my master made me shut my eyes, and close my mind to all the distractions of the body. That is the true trance, boy, the way to certainty. When you become adept even the words cease and there are only patterns. The numbers have no names or places, only connections.

That is the pattern which I've given my life to bringing into the world. There is only one material of which to make it, so that it will last until the world ends.

Stone.

We wandered a great distance over the years. It's possible to travel a long way east, inland, in that country, because the bones of the land are made of chalk, and form ridges of high ground above the forests, where the land is open and one can walk easily. We came to the place where the tribes meet, and there I saw for the first time the greatest circle in the world. It's built on chalk, out of stones that were brought over the sea from the west. I met men who saw it built, though by now I think those days are out of living memory.

I saw how the circle was made – not just the engineering of it, you understand, but the principle, the pattern that came out of the dark place behind the eyes, the pattern that reflects the sun and moon and the movement of the circles that surround the world. I saw the sacrifice. I saw how a man who had the pattern in his thoughts could unite with it in his body, how he could become the pattern himself, and remain so until the end of the world. I saw that certainty was possible, but that there was one way to reach the end. A hard and terrifying road, it seemed then, because I was young and loved life and feared pain. But I knew even as I watched in terror that this was how it would have to be.

The time came for my master to send me away. I had the patterns; it was time to study the stones. I turned my attention from the sky to the earth, and I travelled over sea and land and I came into contact with every kind of rock: granites, sandstones, chalks, limestones, slates, schists. It was hard at first to fix my mind upon what I could see and touch. There seemed to be no meaning in it, only random changes that had no explanation. Then I began to see where order lay; in fact I had always known it but in my wandering had forgotten. There are days and nights and seasons; it is time that gives a circle to the earth. Time and the thoughts of men. We make circles on the land to give it a shape and a purpose. That's the true magic, the mirroring of the patterns of the sky in the material of the earth.

That's what I was born for. It's not the destiny of every man. There is a rule in numbers: one is greater than a hundred, and a hundred greater than a thousand. You see the principle mirrored in every tribe. There's one chieftain, and a few leading men. Then there are widening circles: landless warriors, women, labourers, children, slaves, cattle and foreigners. That is how order is made. My brother is a chieftain, and he killed four of my brothers because it was necessary that he should be one and not divided. He did not kill me, because I stand at the centre of a circle of my own, which is not a threat to his, but the complement to it.

Why should I tell you more about myself? You shouldn't even want to know. I'm a magician and a builder. I was sent north when people from my country first established the colony here. It was a long voyage, always keeping the coast in sight to the east of us. We passed mountains and forests, and just occasionally, at the mouths of rivers, we could see smoke rising which meant there was a village. At last we reached this island, and as soon as I came ashore I felt the land greet me, and I knew that this was the place ordained for my great work. There were circles here already, rightly aligned, but made of wood. People think wood is permanent because it lasts beyond the span of a human life. But what does that mean? The stars move much more slowly than that, and the stones speak in the same time as the stars move. If you could stand still for ten times a hundred years you might stay long enough to hear the story of the stones. That's why the only way to bring the light of the stars down to earth is to reflect their patterns in enduring stone.

This is a land where stone is infinitely variable. I've explored every corner of it. In the mountains at the north end of the island, where nobody lives, there's a peculiar coarse granite, and great round boulders are scattered in readiness across the hillsides. On the coast there are slabs of red sandstone waiting to be cut out of the live rock. I've used both in my circles. We lost more slaves bringing down the granite than in any other way. It took a long time, as common people count it, but for the stones human labour and suffering are as ephemeral as the calls of birds. It was expensive, but the kings knew how much it mattered, and they recognised me as Master.

I chose the site, here in the south-west of the island, with routes to the sea on three sides. When I came the villages were on the coast. It was my circles that drew men inland, and now the circles are in the middle of the settlement with dwelling huts and barley fields all round.

I found the stones. The way I choose is to sit with my back against the rock, so my spine is right against the stone. Then I shut my eyes and in the same dark in which my Master first enlightened me I become stone. Cold and pain are the frail sensations of a moment, part of the babble that common folk call life. I stay with each stone for a long time, and I wait for it to speak. When the voice comes I listen intently. It is the stones that describe the circle to me, because they have always known how it will be. They have been waiting for me, scattered across the hillsides, since the beginning of time. I give them order, and they give me substance. A man and a stone are two very different things, but if they could become one, you understand, that one would be a star on earth.

I've made the island mine. Not as chieftain or warrior: none of that is important to me. Chieftains rise and die again, often very quickly. To seek fame that way is actually to court oblivion. My work will be here until time ends. It will become part of the island, and it will seem to the generations to come that it has been here from the very beginning.

I want you to leave me now. Don't look like that; this isn't death. For the common people the body dies and that's the end. But I've spent my life making substance out of my thoughts, and I dare not risk losing it any longer. If you follow my path, do you realise the risk that each day of life will bring? Have you thought about the tension under which you'll have to live? As a builder, a Master in Stone, you'll be called upon to cross many seas, and yet at sea you'll fear the most, for if you drown, all the patterns of your thought will be dissolved in a waste of water. You'll always be afraid of attack in the night, war, famine, disease: all those things. Not because you're afraid to die, far from it, but because you risk everything each moment that you live, by risking the wrong sort of death.

Tomorrow I shall be certain at last. They will come just before dawn and lead me out to the circle which is my thoughts made into stone. They will strip me naked and lay me over the great stone of sacrifice in the middle of the circle. The midsummer sun will rise tomorrow precisely between the two great sandstone slabs, and a shaft of its first light will strike me where I lie bent back across the stone. The knife must strike at the same moment as the sun and not falter. They will cut my ribs open, and plunge their hands into my chest and rip the heart out of my body. They will hold it high so that the new light falls on it, and the last thing I will see with these eyes is that the thing is done.

I see you shudder. I've dreaded pain too. If I allowed my senses to rule me I could still fear it. But don't you see, my son, if you follow my road there's no other ending? If you go on making patterns out of your thoughts, if you pursue this dangerous magic, you'll have no choice. You won't want your thoughts to be imprisoned in a dying body, or to disappear, turning to vapour with your last breath. No, you'll want them to stand until the end of the world, and yourself to be embodied in them. Your magic becomes your curse, you see. It will dog you all your life, until you can bear it no longer. Then you will choose, and one morning the sun will rise and strike you with its first light, and you will be turned to stone at last.

from Dripping Forest

Mikhail Prishvin

THE EYES OF THE EARTH

Towards evening the wind dropped so completely not a leaf stirred on the birch. More and more people were going somewhere along the road running at the foot of Gremyach Hill. On a sandy path off the road I saw a child's footprint, so small and sweet that I could have kissed it if there was no one to see and laugh.

The talk of the people driving along in the carts resounded from the still water, was clearly heard on Gremyach Hill. A foal ran beside practically each cart. Snatches of the peasants' talk reached my ears: they had finished planting potatoes, some Dmitri Pavlov or other had lost his wife and married again before the forty days were out, but there was nothing else for it, left with six young children on his hands. And Maria had married Yakov Grigoryez, she forty and he sixty, and she owned a heifer, too. People in the last cart didn't catch what it was that Maria owned, and the word passed down the whole train – *heif-fer*!

The air was so still now that you could clearly hear the bitterns' bellows seven versts away.

And afterwards, when a peasant woman went down to the loch with her little boy to rinse her wash and the boy lifted up his shirt to pee in the water, the mother's words carried so clearly it was as if she were standing beside us.

'What are you doing, for shame? Pissing in your mother's eyes...'
So the loch was to her the eyes of Mother Earth? As usual, in such cases, I asked my landlady what she thought about it.

'The eyes of mother earth, naturally,' she replied. 'And mothers in general too: if a woman's eyes get inflamed, folk in the village say that her child must have pissed in water.'

So of the ancient cult only superstition remains, while the poetic image of the eyes of Mother Earth is relegated to the loftier spheres of culture.

It was impossible to fall asleep on this scented night, and the eyes of Mother Earth never closed.

A FOREST CEMETERY

A strip of forest had been felled for firewood, but for some reason the timbers weren't carted off, just left lying there – and in places you couldn't see them at all among the young aspen grove with huge pale-green leaves and the thickly growing firs. For those who understand the life of the forest there is nothing more interesting than such clearings, because the forest is a closed book while a clearing is a page opened. After the pines were felled, the sunlight poured in and encouraged grasses to grow to such a giant state that they left no room for pine and fir to sprout. Aspens, however, overcome even grass and they took root here in spite of the odds – and showed a vigorous growth, like unkempt, flop-eared youngsters. And once the aspens have smothered the grass, the shade-loving firs begin to grow among them, clinging close, and that is why firs usually replace pines in a clearing like this. But here was mixed forest, and the most important thing of all was that moss patches which had been suffocated revived and looked bright and cheerful.

The whole life of the forest in its varied forms could be read in this clearing. There was moss with blue and red berries, red moss and green moss, patterned in minute or large stars – here and there you saw clumps of white Iceland moss dotted with red bilberries.

Everywhere, young pines, firs and birches had sprung up round the old stumps and looked radiantly green in contrast. This vigour inspired happy hopes, and the black stumps, which had once been tall trees, were not a depressing graveyard sight.

Trees die in different ways. For example, the birch rots from inside, and for a long time you mistake its white trunk for a living tree. But inside there's only rotten wood. The trunk is like a saturated sponge and quite heavy; if a tree like that is shoved, the top pieces will fall so heavily, they might even kill the person standing underneath. You often see a birch stump, standing there like a pot for flowers: only the waterproof bark remains white and imperious to decay, while in among the rot, flowers and saplings sprout. But when the firs and pines die, they shed their bark first. It falls from them like discarded clothes, and piles up under the tree. Next, the crown falls, then the branches go, and finally the stump itself disintegrates.

A mass of flowers, of mushrooms and of ferns, is quick to compensate for the loss of a great tree. But, more important still, the tree itself continues to live in that sapling growing close to the stump. For its part, the bright-green moss extends its stars, with their tiny brown hammers, hurries to cover the dead tree's bare trunk which had once held it upright in the ground – very often you find red russulas as large a saucer, growing in this moss. Pale-green ferns, red strawberries, bilberries and bog whortleberries, crowd round the ruins to hide them from view. Sometimes, a cranberry has to creep over the dead stump for some purpose of its own, and blood red berries and tiny leaves add their further touch of beauty to the ruins.

THE GIRL AND THE BIRCHES

The leaves were only just appearing on the birches, and the woods seemed so huge, so unspoilt. I never thought the train a monster threatening them, indeed, I found it very pleasant, sitting at the window, feasting eyes on the endless scope of these laceworked birch forests. At the next window stood a girl, young but plain. Now and then she would glance warily around, like a bird dreading the hawk was watching. And then she'd poke her head right out of the window.

I wanted to see what she was like in that moment of freedom, alone with the birches, and so I got up quietly and peeped out of the other window. She was gazing rapt at radiant young leaves, smiling, whispering something to herself as her cheeks glowed.

THE NEW MOON

Sky's clear. Sunrise gorgeous in the still. The frost's 12 below zero.

Golden light in the forest all day, and when sunset came, a glow spread across half the sky. It was a northern glow, a sparkling raspberry red, the colour of the shiny transparent paper Christmas crackers are made of. But the sky wasn't all red – in the middle of it ran a deep-blue lancet-shaped streak, as though it marked the passage of a zeppelin, and along the edge there were lines of various hues in the subtlest shadings, complementing the main colours.

Sundown's glow lasts fifteen minutes at best. The new moon hung in the blue facing the red glow, and it looked surprised, as if it had never seen anything quite like it before.

trans. R.A. Jamieson after Olga Shartze

TRANSLATOR'S NOTE

MIKHAIL PRISHVIN (1873-1954) grew up in Orel Gubernia in central Russia. He entered Riga Polytechnic, but was arrested as a member of a member of a revolutionary circle and sent into exile. He enrolled Leipzig University in 1900 to study agronomy, where he met and fell in love with Varvara Izmalkova, a Russian student girl – the 'Phacelia' of his prose-poem of that name. Back in Moscow, working at the Academy of Agriculture, Prishvin published his first books in that specialised field. But he lost interest in this study, left the Academy and attempted to earn his living as a writer – 'I wanted to be a real writer and not a hack.'

Contact with a group of ethnographers led to an expedition into the little explored Russian north to gather folklore, out of which came his first broadly successful book, *In the Country of Fearless Birds*. He was admitted into the Russian Geographic Society and a further trip into the far north followed. Between 1912 and 1914, a three volume collection of Prishvin's early stories and sketches was published by the Znanie house as part of Maxim Gorky's sponsorship of new Russian writing, and Prishvin's reputation was secured. After the October revolution, he embraced the challenge of the future, and gladly taught in a school in Yelets, and later as village librarian and schoolmaster in Smolensk Gulbernia.

Here Prishvin's inspiration continued to be the natural world. He was also a keen hunter, an activity which he recorded in various hunting stories. He liked to remind those who questioned the correlation of his feeling for the souls of creatures with killing, that many hunters had made considerable achievements in the field of conservation and human understanding of the wilderness. He wrote: 'If I didn't know the neat distinction between those times when a hare may be shot, and when one should risk one's life rescuing it, I would give up hunting and take up arms against hunters.'

Prishvin gradually turned from nature, to Russia as the Motherland, as his primary theme. In the epilogue to the story 'Ginseng' (1933) he writes that 'Together with everyone working in our new culture I feel that the root of life has come into our creative life from its natural taiga environment, and that there is a better chance of finding it in our art, science and useful activity than in the depths of the taiga.' During the Second World War, though he wrote fiercely against Russia's enemies, his work betrayed a growing misanthropy and hatred of civilisation generally.

Mikhail Prishvin's miniatures remain among Russians' favourite writing. The intriguing mix of poetic with scientific sensibility appeared quite new in Russia at the time, the artist 'seeing life without himself in it, seeing those firs and birches on their own ... Seeing them in their on-their-ownness is the artist's supreme achievement...' As Professor Valentin Lazarev has written, Mikhail Prishvin 'probes the depths of Russian history and the Russian soul itself.'

Andrew's Teeth and the Nature of Reality

Angus Dunn

IS THERE AFTER all, any reality that we can grasp, that we can wholly rely on? Perhaps this world is a phantasm through which we drift, with no choice and no real ability to act. No-one can be certain.

Yet I believe that the matter of Andrew's teeth may give me the necessary psychological leverage to crack open that nut.

Amongst a fairly varied selection of employments, I once spent some time working in an abattoir in Munich. The job was not arduous, as my responsibility was to slip the membraneous tubes onto the dozen or so stainless steel spouts through which was extruded the macerated ends of meat, bones, eyeballs, hooves and suchlike. The tubes, filled with this meatish stuff, were taken on a conveyor belt to the tying station where Donna, an Irish girl, operated the machine that tied them off into fat eight-inch sections.

The place was noisy, but Donna and I managed to develop a wordless relationship during our ten-hour shifts. I was aware of her presence, and I believe that she was aware of mine. When the day was over, and we had removed our rubber boots, ear muffs, eye masks, gloves and overalls, we would sometimes walk back to the hostel together.

This hostel was solely for foreign workers and the common-room was always crowded and noisy with French, Italian and English conversations. Donna and I would talk there sometimes, but she had other friends in the hostel and our relationship was always a possible one, rather than probable.

The rooms in this hostel each had three or four bunk beds, and there was a great deal of indiscriminate visiting in the daytime, when off-shift. Consequently Donna was not surprised when someone came into her room while she was napping. She turned over to look, and saw this person leaving the room. Only later did she find that her silver charm bracelet had gone missing.

We all commiserated with her, and wondered who the thief could have been. I asked if she could recognise the person in her room.

'No,' she replied, 'but I did see that he wore a striped polo-neck sweater.'

Jackie looked at me. 'You've got a striped polo-neck sweater.'

And it was true, I did have such a sweater, but of course I hadn't stolen the bracelet.

'Give us it back,' said Donna. Her smile was small and uncertain.

I shrugged and smiled back. She wasn't really serious and nothing further happened. Donna resigned herself to the loss and I ... well I worried. After all, I could not verify the fact that I hadn't stolen the bracelet. If I had been someone else I would have suspected me.

That night I lay in my bunk unable to sleep, worrying about the nature of memory. I could find no memory of stealing the bracelet, but was my lack of memory proof of my innocence? From ten years before I recalled my father, of an evening, reading Freud's *Psychopathology of Everyday Life* out loud. Convincing evidence indeed, of the power and complexity of unconscious motivations. I wondered whether a mind so devious as that which Freud depicted could make me steal something precious from a young woman I desired – and then make me forget it. I believed that it could.

At 2 a.m. I got out of bed, careful not to disturb my room-mates, and quietly searched my locker and my bags. I was only slightly relieved to find no bracelet. After all, I might have forgotten where I'd hidden it.

I went back to bed and managed to sleep.

Unable to prove my innocence to myself, I spent the rest of my stay there in a state of some unease. What particularly upset me was that several other people in the hostel could equally well have been the culprit, but they showed no signs of guilt. There was no trace of worry in their faces.

I have come to understand that there are some people who are completely confident about the nature of the world. They are certain about the events around them. They can specify precisely the words of a conversation some days or weeks in the past, or the disposition of the various vehicles in a minor traffic accident.

I find, on the other hand, that facts are the most nebulous things and it is a constant amazement to me that a court of law can pass judgement on even such a simple matter as a petty theft from a newsagent's. To accept only one of many possible interpretations of events! I imagine sometimes what it must be like, to have such a mind. At least, I try to imagine it, and shudder.

Not that criminal incidents are the only events about which I am uneasy. By no means. I am not obsessed with concerns of legality. But such incidents occurring nearby do tend to affect me badly. It is always a relief when any local crime is solved, since I can then be sure that I am not the culprit. For similar reasons, I keep a diary so that I can at least give myself some comfort in the case of the more outstanding of national crimes, by being able to verify my whereabouts at the time in question. Not that there is anything but thin comfort available, as I am well aware that the journal could be a complete fabrication. I find it painfully uncomfortable to remain in any locality where there is a high rate of unsolved crime, and consequently I change jobs often.

I have read a great deal, in bits and pieces: I am not uneducated. I have been advised both by professionals and by well-meaning amateurs. I tend to agree with the suggestions that what they call my problem is due to an incomplete process of primary socialization.

I was an only child in an isolated house, where everything was explained to me, many times. The rules of the house, the way to behave at table, the times when I was allowed to speak, the much longer times when I was required to be silent. One effect of all this was that I never learnt to infer the meanings in the behaviour of others. The meaning of an act lay in the rules which allowed it, or compelled it, rather than in any consequence of that act.

Largely this was due to my father, now long gone. I am certain of his role at least, though my mother's acquiescence was perhaps equally important. He was a lapsed minister of the church, and had adopted an idiosyncratic form of logical positivism to fill the gap left when faith departed. If someone threw a towel on the fire, turned the radio on full-blast in the night and scuttled back to bed, or chopped down the cherry tree, as it might be, the evidence logically suggested that I was the culprit. I remember, before I finally gave in to the authority of his black leather belt, trying to present my version of events, but he was positive in his assertions that I was the guilty party.

When it came to proof, my story was always ignored, while circumstantial evidence was given the status of empirical fact. The matches were found in my bed, so I lit the fire. My version of events was always outvoted by my father and my acquiescent mother.

I recall that I once did commit some forbidden act, by negligence, rather than by bad intention. In something of a panic at the thought of a beating from my father, I carefully prepared the evidence so that it proved that my mother had done the deed. The logical inference of my mother's guilt was completely ignored. I smart still at the injustice of the punishment I received.

The effect of this childhood training was to lead me to doubt my own testimony, and hence to doubt my memories. I cannot now honestly say that I did not perform the other acts for which I was punished. The most that I can say about these long-ago events is that I may have been responsible.

However, I have long since come to suspect that my father had a hand in at least some of them.

When my mother was dying, I went to see her. I tried to induce her to talk about my childhood, but she was only interested in recriminations.

'It nearly broke your father's heart,' she said, 'when you cut down the cherry tree.'

'But it's only a figure of speech, mother. We didn't have a cherry tree.'

'No. Not after that.'

'But we didn't ever … '

She looked at me reproachfully. 'It's just as well your father's gone before. The poor gentle soul would be shocked at you. Telling such fibs to your mother.'

'The poor gentle soul? But he wasn't … '

'That's enough. I'm tired now. I think I'll have a little sleep. You can come and see me tomorrow.'

She died that night.

This is the way it is, for me. I live in a state of constant agony, uncertain of everything, foundering in the meaningless flux of events. The world is as a running stream in which the facts of life are eddies and waves that disappear in a moment.

I have lived in many places and in many different ways, and sometimes I have heard people extolling the virtues of such a viewpoint. I can only assume that they are not afflicted with this unfortunate state.

In addition to the socially reprehensible lack of any opinion on anything, I have an unfortunate tendency, when psychologically distressed, to take a stand on the most ridiculously untenable positions, purely for the sake of being, at least temporarily, sure of something. This certainty lasts only as long as it takes for the next person to start speaking. Even if, as has happened more than once, this person agrees with me, I instantly lose any sense of confidence in my own argument.

Yet it seems to me that I have caught a glimpse of a process at work that may help me to find a touchstone. In the external world, I cannot verify my experiences by audio or video recording: this would be a cumbersome process and would attract attention. Similarly, soliciting affidavits from witnesses would be considered odd. But I believe that I may have some chance of approaching verification of reality from the other direction, by outwitting the mental processes, by getting in behind them to see what is really there.

Probably it is an everyday event for most of us to censor our memories or to adjust the parameters of memory so that our version of events is more flattering. And there is a similar process at work which enables us to modify our memories to match those of our peers, our friends, our colleagues. There is a great deal of social pressure on us to conform, and this is not limited to dress codes and manner of speech.

A couple of years back I had a job on a building site with Andrew, a brickie. He used to bring his dog to work, an Irish Springer Spaniel. One fine day, sitting in the lee of a shed to eat our lunch, Andrew took his teeth out.

'Watch this,' he said. 'Here Tash!' he called to his dog.

The spaniel ran over and sat eagerly in front of him. He grabbed her by the head and thrust his teeth into her mouth. She looked very comical and we both laughed as she threw her head about, until she managed to shake the teeth back out.

Still grinning, Andrew picked up the teeth, wiped them cursorily on his coat sleeve and put them back in his mouth.

Astonished as I was, I said nothing. Though I tend to find people's thoughts and motivations difficult to divine, I have gathered that there is a certain social finesse about such matters, which varies with occupations. I had not worked in the building trade before. Perhaps among brickies and masons, this behaviour was considered normal.

I ran into Andrew from time to time, after the job had finished, and I met him in a bar a couple of years later, to celebrate his engagement. During the evening, various acquaintances told tales of his terrible past, as the way is, to embarrass him in front of his fiancee. I took the opportunity to join in.

'Do you mind the time you put your false teeth in the dog's mouth? On that job over at Gairloch?'

'I never did! You're making it up!'

And in that moment it happened and I saw it. In the face of his obvious desire not to be associated with the act which I remembered, I let the matter drop. But also, mirroring this adjustment in social reality, I perceived, fast as a wink, a matching adjustment in my own internal reality. Somewhere between two thoughts, something flopped, a gate closed, and what had been a sure and certain memory was suddenly fogged and dim. An adjustment had been made in the status of the memory.

So now I know how he works, the villain of the piece. If I didn't exactly see his hand at work, I saw the shadow of it.

I'm waiting now, and next time there is a sudden deletion or an adjustment of memory, I'll be in there like a shot, in behind the thoughts and false memories to where reality is.

Angus Dunn is currently writer-in-residence with Aberdeenshire Council. He lives at Aroch in Ross-shire and is an editor of *Northwords* Magazine.

Far Verden, Farvel

from the novel *Barbara*

Jørgen-Frantz Jacobsen

THE KIRK BELLS OUT AT REYN announced the service. They had a mild tone that fluttered over the house roofs. But the tower where they hung was gloomy and broken down with age. It trembled under the motion. The bells hauled and ground at the wood, and a dismal creaking blended with the ringing birdlike song, to be heard over all the town.

The storm and the rain had stiffened into still coldness. Dirt in the street had become hard and sharp, and water puddles frosted tinkling ice. The sun shone wanly over the twenty-sixth Sunday after Trinity.

The kirk was small. There were thirteen pews on either side and a little gallery over the entrance. It was ice-cold between the bare wood walls. There was no loft, and eyes looked straight up to the rafters and spars.

Torshavn's poor folk were beginning to appear. The women wore black shawls and headscarves. They came in tight-mouthed, Kingo's psalmbook in hand. Their men followed after, hesitating a little, with a curious desire to remain in the background. A few wanderers among them found their way up into the gallery. They were everyday heroes, Samuel o' Vippen and Niels-in-Puntin. Here in this holy place, it was better that the women go before them.

And so came the better class of folk to kirk, and the *fine*. They had their set places, that was their right, which none of the plain folk dared challenge. Bailiff Harme came and spread himself out beside his daughter Suzanne. They sat in the foremost rank. There was also the Magistrate's Bench, which was still being used by the widow of the late magistrate Mme. Stenderup, and the Priest's Bench, where the two sisters Armgard and Ellen Katrine had their place with the Reverend Wenzel's relatives and guests. And the Reverend Poul Aggersøe found a space here too.

There were also many others who had their fixed seats though they were not as prominent – both Mme. Dreyer and Sieur

Arentsen, and the old spinster Kleyn. These were stolid folk who paid for their pews. It was a matter of pride that marked them out from the crowd, and could be traced faintly on their faces, as they sat themselves down. But common to all the churchgoers was their breath, clearly visible as grey clouds in the bitterly cold air.

The Reverend Poul sat and thought of all the times he had gone to the Church of Our Lady in Copenhagen, of the whole elegant tumult outside the gates, of coaches and chaises, and lackeys who slammed carriage doors, of the spruced-up throng of distinguished and ordinary citizens, who strode, smug and dignified between the pillars, through the high naves, while the organ sent the sombre prelude howling through the arches. Ach, here in Havn there was only a poor bent clerk, who spluttered hoarsely through the opening prayer. And all the time it was dead still under the rafters, no coughing or croaking as was usual.

But just then Barbara came in hurrying softly and sat gently down on the Magistrate's Bench next to her mother – a last living glimpse of the world before the solemn hymn began.

That day Magistrate Johan Hendrik Heyde had come to kirk along with the Bailiff and his daughter. He'd greeted them, exchanged a few words and smiled, but when they stepped up into the porch he hadn't gone with them up the aisle to his bench. Instead he traipsed up onto the gallery.

Why did he do this? He had nothing against them, but at that moment in time there was one matter that marked him apart from them. They spoke Danish. Fair enough, that was their language. And Johan Frederick had nothing against Danishness, he had often stayed in Denmark himself, his own family was Danish. Yes, it would be absurd if people like the Bailiff or the Manager of the Trading Company spoke Faroese. And yet, here on home ground among the ordinary Faroese folk, this slick tongue broke the mood. It annoyed him a bit, like when he heard a musician playing flat. He himself was attuned to the folk. So he objected to walking forward to the foremost benches. He preferred to stay in the background up in the gallery. Here he could sit in peace and think over agricultural improvements, new experiments in fishing and other subjects useful to the country's wellbeing.

Some time after the Magistrate had sat down, the stairs creaked heavily and up came his cousin Samuel Mikkelsen, the Lawman of the islands. He was big and awkward, but he too preferred the gallery. Not from any subtle aversion to other fine people, but out of discretion. He followed a circumspect path in all things, he told no joke except with sympathy, and when he took a sip during the service, he did so with great delicacy – not furtively like a rogue or a schoolboy, but with imperturbable dignity. Johan Hendrik felt soothed by his near presence, but the Bailiff Harme maintained, not without some grounds, that His Majesty's Faroese-born officials misrepresented their station by skulking among the whingers on the six free benches in the gallery.

The Commandant, Lieutenant Otto Hjørring, was different. He didn't hide his light under a bushel but strode into the kirk in a tailcoat, with sword, handlebar mousetache, pigtail, looking every inch a military personage. But despite all this, he failed to carry it off when he stumbled into one of the commoners' benches. Tangloppen the Sandflea was both honoured and anxious. He gawked repeatedly towards this overpowering proximity of glorious colour and delightful scent.

So they had assembled, all the folk of the parish, high and low. Gabriel in his Sunday best, pious and unlike himself. Employees of the Company and the officers of Skansen, the farmer of Husagardur with his servant, and the farmer from Sund, who had come a long way on foot. The congregation laboured through the first long hymn. There was no organ. The frosty trembling voices slipped out of tempo with each other. Some sang out well, following the lines, like Sieur Andersen, who obviously considered himself the leader of the choir. Others sang as if they hadn't a tone to their lives, and the women only wailed up from their black headscarves. There came a tremendous coughing and snorting as the last interminable, heartfelt verse sang out to its conclusion. The model of the East Indiaman 'Norse Lion' that hung in the rafters was turning slowing on its mooring. Its bowsprit came to be pointing to the south.

The minister, who had been standing faced to the altar, turned towards the congregation: – *The Lord be with you!*

A steam cloud blew out of his mouth. A hundred damp clouds from the congregation answered: – *And with your spirit!*

The Reverend Wenzel was a little too small for the red chasuble, and it seemed like he might trip in the folds of the surplice and fall at any moment. He threw a glance over the assembly. The Manager of the Trading Company hadn't taken the trouble to come. Shameful. No, that man was a sadly mediocre churchgoer. Today it was especially annoying. The minister was disappointed. The sermon of the day would hopefully have a message for one and all, but most likely it was for the great and worldly, who forgot the kirk and overlooked its workers easier than most people. The Bailiff was there. But as the Reverend Wenzel turned again to face the altar, he saw that his own wife wasn't. Anna Sophie hadn't come yet!

– *Let us all pray!*

Vague disquiet began to fill him. He fell into a daydream in the middle of the ritual. So he swallowed hard and chanted the collect through with the faintest warbling in the voice. No one found it strange. The Reverend Wenzel often spoke as though he was upset about something or other. The whole of his small red-bearded frame always seemed a little peeved.

The service took its course. It reached the epistle. The Reverend Wenzel, feeling oddly abstracted, pulled himself together. He turned towards the congregation and intoned in a strong voice: – *The Epistle on the twenty-sixth Sunday after Trinity is written by St Paul in his first letter to the Thessalonians.*

Anna Sophie hadn't come.

– *And we beseech you, brethren, to know them which labour among you, and are over you in the Lord, and admonish you; And to esteem them very highly in love for their work's sake. And be at peace among yourselves. Now we exhort you, brethren, warn them that are unruly ...*

For quite a while after he finished the epistle, the Reverend Wenzel still heard the echoes of his own voice in his inner ear. His heart was uneasy, his brain vacant. He hadn't connected meaning with one word of what he'd said. Nor had the Reverend Poul, sitting on the minister's bench. His restless mind wandered down other roads too. Yes, in the whole kirk, hardly anyone paid attention to the apostle's words. Some were too worried, others too sleepy. And so the service went on like a game, once started. The 'Norse Lion' had swung imperceptibly and was now heading sou-west.

As soon as the minister mounted the pulpit and started to preach, Johan Hendrik up the gallery could tell there was something eating at him. That peculiar tone of voice he knew so well. He had known it since they were boys.

The Reverend Wenzel began tentatively, but soon gripped the sure thread he had laid out, and followed it quickly. This was the sermon he had prepared so diligently, the one the Manager of the Store should have heard, even though it contained as much of a message for the Bailiff, yes, actually when it came down to it, for the whole congregation right down to the lowest of the low.

– *And we beseech you, brethren, to know them which labour among you, and are over you in the Lord ...*

It wasn't every year there was a twenty-sixth Sunday after Trinity, and it wasn't every minister who was permitted to preach on the epistle rather than the gospel at the morning service. But the Reverend Wenzel, who was a magister, was allowed to do so, and at last, on this rare Sunday, he had the chance to vent the contents of heart.

Anna Sophie still hadn't come.

She too might have benefitted from hearing his words. Although it might have run like water off a duck's back. No, it probably wasn't because of this that he missed her. He didn't allow his mind to dwell on it. But his heart knew better, it pumped violently.

He began by rendering unto Caesar those things that were Caesar's. The King's officials and clerks and all the secular authorities – *Everyone should respect and obey them. Because these authorities were instated by God! But – how was it with the servants of the Word of God? Should they not be honoured and respected in the same way? What was their office? The apostle St Paul explained it in the letter to the Ephesians. Give now heed to yourselves, he writes to the principals of the Church, and to the whole flock among which the Holy Spirit instated you as bishops to foster God's Church, which he won with his own blood.*

He looked around him for the first time and raised his voice: – *But when bishops and ministers give you your spiritual food, and guard your souls, and justify your ways to God, should they not then be considered equal to those – yes at least equal to those – who feed you corporally, look after the supplies of the country* (the Trading Company Manager should have heard this!), *look after your worldly welfare and keep law and order and justice among you?*

The Reverend Wenzel had warmed to his talk: – *Spiritual and worldly are different, but among these the spiritual is not inferior to the worldly. When you keep the laws of the world, as the authorities maintain them, should you not also keep God's eternal commandments, as I preach them?*

He looked appealingly at the Bailiff, and Bailiff Harme nodded.

– *And when you pay rent on your land and taxes to the King's clerk, should you not then also pay your tithes to God's Church?*

The Bailiff again nodded his agreement. The Reverend Wenzel swung himself higher: – *And when you give service and pay the land tax to the military and its officers, are your principals in the Lord not then worth so much, that you should give them tithes?*

His eyes strayed to the Lieutenant, who was fast asleep in all his crimson glory. Then his glance returned to the Bailiff, yes, it was almost as if this sermon was going to turn into a dialogue with the Bailiff. And the Bailiff again nodded graciously.

– *Because the minister*, the Reverend Wenzel continued: – *The minister is your spiritual superior, instated by the Holy Spirit – and indeed also by His Majesty the King – in the same way as the Bailiff and the Lawman, the Manager of the Royal Trading Monopoly and the Magistrate …*

He sent his brother a look, behind which lay a kind of mild reproach. But then he glanced again at the Bailiff and went on: – *Eh, in the same way as these are your high worldly superiors, whom you should respect and love. Do not take umbrage with the minster, because he tells you the truth. Even the highest should not let his admonitions pass them by unheard, if not for his humble person, then for the sake of his office.*

Now it was said. The Reverend Wenzel paused, his waterblue eyes wandered a little uncertainly, as if he sought to discover the effect of his words. Then he carried on: – *Even today the apostle speaks to us: But we beseech you brethren, chastise the ungodly ...*

Suddenly all was lost. Anna Sophie – now he saw her! And he saw the Manager of the Trading Company too. Out through one of the kirk windows he looked across to the manse, and there, in the bedroom, he saw both his wife and the Manager ...

He stood swaying, understanding nothing, feeling nothing. It was like a dream. Anna Sophie and the Manager of the Royal Trading Monopoly! How could this be ... he struggled to realise it ... but there had always been something, yes, he'd always suspected!

He wanted to go on from where he had halted, but a stinging pain grew in his heart. A paralysing numbness broke loose inside him. His heart tightened into an ironhard knot, it squeezed up and burned through his throat, it radiated pain through his chest.

Anna Sophie!

He heard his own voice, still stammering out: – *Chastise the ungodly ... chastise the ungodly ...*

The whole congregation saw him, how he stood reeling, helpless, his face chalky-white, with a little fire-red spot on each cheek. Some of them thought he had fallen ill, but no one knew the reason for his distress. Quite a while passed; the lieutenant's steady snoring became irregular, he muttered some words in his sleep.

Then it seemed that the minister himself noticed the great silence. He realised he was standing in the pulpit, not preaching, while the congregation sat in the kirk, not listening. A glimmer of sunlight fell through the windows, he saw all the faces clearly. At the top, in the background, his brother sat with a puzzled expression. The 'Norse Lion' swayed slowly on its cord, aimlessly. And the kirk was like a ship, drifting masterless among the waves, full of folk, but without a helmsman.

The Reverend Wenzel collected himself. The ungodly – now he had it! He let his voice sound out, it trembled with injustice. If he had an impassioned cause before, he had no less of one now. The ungodly – who else but the mighty? He visualized the Manager's smug, smooth air. Lightning quick it was clear. He had admired this air, tended it like a plant when the Manager sat at his table over a fine steak. Now he saw it was his own happiness that was being devoured, and he himself trampled upon by that same self-regarding, matter of fact manner. He gasped, his body and soul turned away from the mighty, he snarled at them like a beaten dog who had done no wrong. He suddenly imagined them as an immortal mob of potency and smugness. The Bailiff, and behind him a row of bailiffs, who despised the farming folk. The Lieutenant, and behind him a row of lieutenants, who bullwhipped their orders into decent men. The Manager of the Trading Company, and behind him a rabble of managers, who had cheated his worthy, naive forefathers. Did he not know about the great and the mighty? King David, who stole Uriah's wife. Herod, who had John the Baptist beheaded. And behind them all, Pilate, who washed his hands. The great always washed their idle hands as a last resort. Did he not know them? They had been the worm of his life. Just a few moments ago, he had stood there aspiring to be one of them. And now he was stabbed in the heart!

Trembling with bitterness the Reverend Wenzel had begun his reckoning with the ungodly of the world, but as he searched deeper into his soul and saw the beam in his own eye, he swelled with his preaching. No hidden corner of the human heart, no sneaking selfish thought, no earthly desire, no foolish vanity escaped his discovery. He made it plain that the world, and only the world, was the object of the human mind, because all were born bad and sinful and didn't know themselves what great sinners they were. No one could justify themselves by their own power or merit. All lacked honour in the sight of God. Every time a person did a good deed, right afterwards came conceit and complacency, the great flatterer that held out its hand and changed the good into evil. Nothing good was said or done but the world lay beneath it, a dirty ulterior motive, so infinite was man's incapacity for true goodness. At bottom, all was desire, falsehood, self-righteousness, selfishness, folly, vanity! Vanity!

– Hmm, said the Magistrate: – So he's swinging into *that* mood today. Yes, something's being going against him for sure.

But the congregation down in the kirk listened with wonder and dismay to the Reverend Wenzel's sermon and thought that he was, after all, a great preacher and chastiser, because to them it was as if he was showing them their own image in a mirror, and could see deep into their hearts. They didn't realise that it was only his own heart he was looking into.

And so too the new minister to Vagar, the Reverend Poul, was obliged to bend his head with the others. Because he felt that he had wandered along the same path of vanity and was never entirely happy unless his heart was flattered, his ambition excited and his desire awakened. The world it was, and only the world, that flirted with his eye. He didn't have the capacity to lift his gaze higher.

But the Reverend Wenzel, who saw that he had won the congregation's attention, delved down deeper and deeper into the underground labyrinths. When he could get no further, when he had fully revealed man's utter lack of propensity for seeking the good, and just how unworthy he was of ever seeing God, then suddenly he let the miracle happen. And this miracle was the mercy through Jesus Christ, who took all our sins away! Yes, through mercy and mercy alone was it possible to rise, release the burden of the world, despise it for the filth it is, and finally with a true heart turn one's thoughts towards heaven.

– *Amen!* he concluded in a loud voice, and before he began the prayer, his pale eyes flickered round the kirk, without coming too near to a certain window. It was a moment of grace in the parish of Havn.

When they reached the hymn, the Reverend Wenzel announced they would not sing the one prescribed in the psalmbook – he was a magister and could grant exemption – but another, which was well known from 'Spiritual Choruses' or the book 'Thousands', and was called *Tired of the World and Longing for Heaven*.

Most of them knew it off by heart. Sieur Arentzen led the singing elaborately, ably supported by the Reverend Wenzel himself.

> *Travail world – farewell!*
> *I'll never again endure your thrall.*
> *The burdens that on me you heaped up so high,*
> *These I cast from me and henceforth deny,*
> *I'll tear myself loose and win myself free*
>   *From vanity,*
>   *Vanity.*

They were all carried away by the hymn, and the whole congregation sang the next verse with all their might:

> *And what is this all*
> *That the world so haplessly happiness calls?*

Tangloppen shot a sidelong glance at the commandant, who was still fast asleep with a decidedly unmilitary expression on his face.

> *It is mere shadow and drifting of cloud.*
> *It is mere bubbles, bursting so proud.*
> *It is mere thin ice, filth and deceit,*
>   *This vanity,*
>   *Vanity.*

Oh yes, no doubt that was right! Samuel Mikkelsen had a slight headache after last night's boozing. It was the fifth night in a row, and he was slowly beginning to feel disinclined towards the tasses and glasses that were lined up on the dinner table about midnight. No, no earthly pleasure was without its bitterness.

> *What is my age*
> *That stealthily, sneakily maketh passage?*
> *What is my worry? My thoughtfilled mind?*
> *My sorrow? My joy? My forehead so lined?*
> *What is my effort? My labour? My sweat?*
> > *But vanity,*
> > *Vanity.*

The young didn't believe it. But there wasn't a lined face in the kirk that didn't become thoughtful at these words. From Bailiff Harme with all his great responsibility, to Samuel o' Vippen with his eight children. The first had to admit that his twenty-four years of office was merely writing in the sand, while the second only wanted for three shovelfuls of earth to mark the worldly conclusion to his carter's life. But to the Magistrate, the perpetually dubious sceptic, came the sour thought that a piece of useful work on earth was more gratifying than ten cries of jubilation in heaven. He read books of new ideas and had come to believe that hard work and patriotism were more blessed than palm-branches.

> *Oh riches and gold,*
> *You idol of earth, false shining mould.*
> *You that are yet the world's basest of things,*
> *That grow and decay and continue to change,*
> *You that are still held in highest esteem,*
> > *By vanity,*
> > *Vanity.*

This comforted many who had lost their worldly possessions. But Gabriel's unbridled thoughts began to wander into deep speculation, whether various sums of money that had lately found their way into his pocket from out of other peoples', shouldn't be invested in four 'gyllin' of land that was up for sale in Mikladalur.

> *Ach, honour so fair,*
> *What is the worth of the coronets you bear?*
> *So often jealousy lurks on your spine,*
> *You're trodden on; rarely, if ever, feel fine.*
> *You often stumble, where others will glide*
> > *On vanity,*
> > *Vanity.*

Who could say themselves free from it? Who didn't have their own worm of ambition, great or small? From the better folk's pews to the benches of the poor, there wasn't a single soul who didn't secretly look askance at a rival or a superior. Envy lurked on all their backs. Who was without pride, who had no hidden wound?

In the Reverend Wenzel's mind, sorrow and a strange salve ran together in huge whirls. The song comforted him. Behold! Thomas Kingo had felt the same as him and survived.

> *Ach, favour and grace!*
> *You hastily rising, then falling breeze.*
> *You cunning and flattering, comforting wind,*
> *Thousands of eyes and still you are blind.*
> *What are you, next to the sun in the sky?*
>   *But vanity*
>   *Vanity.*

What *is* the matter with Wenzel today, the Magistrate thought again. Something must be tormenting the little man right to the bottom of his minister's soul. Had the Bailiff not bothered to listen to his sermon? Something had happened, that was clear, and now he was standing there, Wenzel, God help him, wading through all the earth's passion with his countenance lifted up towards the light of grace – natural enough, pff! This hymn was becoming absurd, like a spirit used to soothe when the good folk burnt their mouths on the gruel of life. As soon as a man lost three marks at cards, then farewell to the world! Yes, and a fat lot of good that would do.

But when they came to the next verse, it was like something snapped momentarily inside the Reverend Wenzel. He forgot about grace and sang out his misery:

> *Ach, faithful friends,*
> *Turning as quick as the weathercock bends,*
> *Your pretty illusions, fortunate rogues,*
> *Fail us so often when drinking of woe.*
> *You are, as I know from experience base,*
>   *Just vanity*
>   *Vanity.*

And aflame with the anger and zeal of the just, he went on:

> *Ach, carnal lust,*
> *Whom many with death's lips have earnestly kissed,*
> *Your kindling tinder, your flaming ember,*
> *Has sent many souls into souls into Hell to remember ...*

The aura around Barbara's head began to glow. Because it was towards her that the thoughts of the majority now turned, in envy, in desire, in condemnation. Gabriel desired her painfully, even the Magistrate up in the gallery sat and watched her. But she sat as untroubled as a bird on a fence, ignorant of the fact that Hell was flaming all around her. The new minister to Vagar, the Reverend Poul Aggersøe, loved and dreaded her.

> *Your toast is like honey, but the drink it is bad,*
>   *Like vanity*
>   *Vanity.*

Armgard was so small and old in her black clothes. She nodded confirmation. To her, this hymn lacked only a verse about alcohol, the devil's brew which had dragged so many down to destruction. But Ellen Katrine stared happily in front of her, overwhelmed by her memories: – The world!

> *Travail, then, farewell,*
> *You shall no longer deceive my soul.*
> *Deceitful world, I now take my leave*
> *And sink you down deep in forgetfulness' grave.*
> *I long for the ease of my sorrow and harm,*
>   *In Abram's arms,*
>   *In Abram's arms.*

And they shook off the yoke of the world's trauchles and sang out the blessedness of heaven. They had settled with their hearts. While the congregation rose and the bells began to sound in the woodwork of the tower, the plaintive sound of the hymn still echoed in their ears.

Vanity, vanity!

The Reverend Poul walked slowly down the aisle. He recalled that Sunday when he thought he had said goodbye to the world in Our Lady of Copenhagen. But the world had come to him. Here there were no pillars or arches, no organ and no vaults, just a low wooden room of a kirk. And yet the world was just as strong and just as difficult to shake off.

The Reverend Wenzel came and took him by the arm, red flushed and somehow not himself. He seemed like an authority from a very high place, humble and shaking with triumphant joy.

– Now then, Johan Hendrik, he asked in the kirk porch: – What did you make of that sermon?

– Aye well, said the Magistrate and took his time in putting his hat on his head: – The end wasn't like the beginning! But despite that, it was like you in both.

But Samuel the Lawman, who had emptied the first bottle of the day by this time, gave a friendly, clear-eyed greeting to his old paternal aunts, Armgard and Ellen Katrine.

It was snowing a little. Outside the ringing of the bells sounded freer, it fluttered among the white snowflakes. They all felt oddly uplifted. They had seen themselves and each other in the tempting mirror of the world and were enchanted and appalled by everything on this earth. But praise the Lord! The world had no power over them, they had torn themselves loose. Relieved and released, they scurried home, not realising that it was anticipation of their Sunday lunch that gave them wings.

Inside, in the empty kirk, the East Indiaman 'Norse Lion' still hung from its cord, with cannons, pennants and flags, swaying slowly through the points of the compass, a while in this direction, a while in that.

Translation by Liv Schei & R. A. Jamieson

TRANSLATORS' NOTE

Born in 1900, Jørgen-Frantz Jacobsen wrote only one novel, yet *Barbara* is recognised as one of the classics of modern Scandinavian literature. A Faroeman, and a cousin of the writer and artist William Heinesen, he was sent to the prestigious Sorø Academy in Copenhagen, and afterwards studied at Copenhagen University. Despite his island origins, Jacobsen never returned to Faroe. His adult life in Denmark was a continual struggle against tuberculosis until his death in March 1938 – before he had completed his novel. In his last letter to his cousin, he wrote:

> And so, my friend, there remain only the two-thirds of the Mykines chapter and the three final chapters. My poor heart beats. It is cold, but spring is in the air and in the light ...

Although both Heinesen and Jacobsen were from bilingual families, *Barbara*, like Heinesen's work, was written in Danish. In this extract, Jacobsen reflects the linguistic tension between the local tongue and that of the state, through the character of the 'improver', Johan Hendrick. Heinesen's rumoured receipt of the Nobel Prize in 1981 revealed the complexity of the situation when he wrote to the Swedish Academy declining the nomination because:

> The Faroese language was once held in little regard – indeed it was suppressed outright. In spite of this the Faroese language has created a great literature, and it would have been reasonable to give the Nobel Prize to an author who writes in Faroese. If it had been given to me, it would have gone to an author who writes in Danish, and in consequence Faroese efforts to create an independent culture would have been dealt a mean blow.

As a 'bestseller', Jørgen-Frantz Jacobsen's masterpiece was not long in finding a Faroese translator itself – one of thirteen languages into which it has now been translated. It is based on the tale of the beautiful femme fatale of Faroese legend, Beinta (in reality Bendte Christina Broberg, 1668-1752) who is reputed to have enchanted three ministers during her lifetime, but Jacobsen's character draws largely on his own great love affair with Estrid, later the wife of one of the early co-directors of Penguin Books, Bill Bannister, who published the first English language version in 1948, translated by Estrid herself.

The author of the hymn in the above extract is Thomas Kingo Hansen (1634-1703), the greatest Danish baroque poet, who was Bishop of Fyn. He was commissioned by Frederick III to compile a new official hymnbook, which wasn't initially approved, but later became enormously popular as 'Kingo's Hymnbook' – 'Kingo' himself was of Scottish descent.

In 1996, a feature film of the story was made in Faroe by the director Nils Malmros, starring the Norwegian actress Anneke von der Lippe in the title role, at a cost of £4 million. Malmros' *Barbara* was released in the autumn of 1997.

Gaelic Lessons with Elgar

Gregor Addison

I SHAVE IN FRONT of the bathroom mirror. The air throttles the faucets, gurgling in the pipes, spluttering out amidst gulps of peaty brown water. It colours the shaving foam an unhealthy yellow. I draw an old razor across the rough skin, resisting the stubble, bunching the skin into folds that cut beneath it. The blood disappears amidst the scum leaving holes in the surface like holes in the ice at Västmanland where I once watched kids drop weighted lines, tugging out monster fish, hungry for light and gulping air.

Every three or four days or so I shave. It's not a task I enjoy but I suffer it by necessity. The sharp sting of alcohol tightens against the skin. The flesh feels younger. I loathe the scent but thole it anyway. It draws the lines out of my face. Momentarily. For an instant I feel young. I should never have grown old.

I have a routine. Rise at eight. Wash. Shave. Put on my jacket and sit at the table, dressed as if to go out, drinking cup after cup of tea. It's a ploy. I have grown so used to my own company that venturing outside has become tinged with uncertainties. And not a touch of paranoia. Where am I to go?

The village has a small grocery, a tourist shop which calls itself, grandiloquently, 'The Gallery', and a post office tucked aside at the edge of the village store which sells very little. The vegetables are best on a Thursday. The fish on a Tuesday. Newspapers arrive only with the afternoon ferry. Weather permitting. Few of the shops are here to provide a real service or any other provision other than comfort for the lonely. They form a focus for the unspoken loneliness that is this place, the swollen silence that wells up at the beginning of each summer, waiting to be burst by the caravan of tourists. I have all manner of ploys designed to put off the moment. It seems I'm an old hand at this.

In the height of summer the water can drop so low that it becomes infested with sea lice. You can pour a glass of water the colour of white wine and, holding it against the light, see the translucent bodies moving within. It might disgust you, but you accept it as one of the consequences of living here. Lowering yourself into a tub of bathwater that is brown and thick with peat is a dubious ordeal but it gives a fine flavour to the cheapest of whiskies. And don't they say that peace is not easily won and carries with it its own price?

I half look at my reflection in the mirror. On the shelf behind me is a soft elk, given to me by a pupil in Västmanland. A leaving present. I call him Elgar. He watches over me, ready to be summoned to arms if the razor cuts too deep. Ania told me that a grown man ought to put such childlike things behind him, but I disagree. Elgar is a comfort for me. I make up stories about him. He now has a fascinating past to bewilder the local children with. His grandfather fought with the partisans in the Norwegian mountains, sneaking over the border, through the forests, from country to country. Elks have their own country. Their borders are internal and patrolled by hunger. They roam where they will. We have a lot in common, Elgar and I. And together we attempt to learn Gaelic in the evenings. I find comfort now in the childish things in life. The childish things I all too readily left behind. Sometimes, we read together also, or, if the evening is damp, we listen to the radio. Mostly we roam.

It was on such an occasion that I came across the little house. It had been a lovely house in its time but now the slates had collapsed, the wooden rafters rotted, the sheep defiled the concrete floors. It sat a little down an embankment of unkept fields where the furrows had collapsed and grass and rushes covered the entire area. The ground under foot was bog which the grass mulched into with the crush of your heel. It was treacherous land. But I made my way, bit by bit, down to the little building.

At the door there was a patch of sunken mud, covered with mirky water, which made entrance through the front door impenetrable, were it not for a large boulder which sat in the midst of it and a heavy piece of wood, a rafter, apparently, which someone had evidently used for the same purpose which I sought to use it. I placed a foot on the wooden ridge and stepped forward, feeling it sink below me, the side of my boot sliding into the mud, but deftly, and with desperate agility, I leapt to the rock, touching it for just an instant, and landed square in the door frame like a crucified Christ, my hands pinned for support against the door jamb.

There were two rooms downstairs and an upstairs which appeared to be more like a loft, though it was certainly two bedrooms, and to which a ladder, broken now and distinctly dodgy, was the only way of access. I pushed open a door to my left and startled some small finches which took off through the broken window, built up somewhat with rocks, as though someone had made a vain attempt at conservation. There was an old brass bed frame, rusting now, and a chain in the fireplace, a *swey* or *slabhraidh*, from which the cauldron might hang, dangling down the chimney a foot or so above the hearth.

Set in the distant wall beside the fireplace was a small alcove. I could see that there were jars and bottles there and so made my way through the excrement and mud to settle my further intrigue. On two shelves were unopened jars of preserves, the tops tight with gauze. Sealed wax, through which a length of string, looped in a noose around the lip of the jar, appeared dirty rusted yellow and brown. The string inside the wax, as I picked a little with my nails, emerged white and pure, untouched by time. Remarkable, I thought. And even more remarkable was the corked bottle of sherry on the upper shelf. Dated 192–, its label worn and darkened by the peaty air, the glass a deep brown, like demerara.

I was overcome with a feeling to carry them away, to preserve them, and horde them away for posterity in better circumstances than these. But I was unable to carry them all. And I didn't doubt that my presence in the community would be felt more bitterly if I was branded a thief. Leaving things as they were I made my way back out of the little house and up the embankment to the road, pausing for a moment in breathless exertion to look over the bay, at the black moving in from the north, pressing through the grey haar to the bright horizon of light beyond. Who were you? Who were you that you stood against all the vagaries of time to leave this mark of presence here? I wiped my boots on the fern at the roadside and crunched the gravel beneath my soles. The afternoon service had skailed and I had only an hour or so until the evening service began. Setting my face against the horizon of light, I made for the headland and the little manse, its windows glinting in the afternoon sun.

The towel is hard and dry. It ought to have been washed about a week ago. I seem not to have been able to summon myself to the task. Time has not availed itself of these few favours. I press it tight now to my face and allow the blood to soak into it. Pushing it up over my tired eyes I mufffle a sigh that almost brings its own tears. The towel is warmed by my breath. It dampens against my skin. In this way I learn to enjoy moments of exasperation, drawing warmth from them, cold comfort. Tears can also be a comfort. But I choke them back. Steady now. I look over the parapet of the towel at the figure in the dark leaded glass. I stay poised like this until I feel I've staunched adequately the flow. Then I remove the towel and tilting back my head swallow air like a fire eater. Like a sword swallower. I let it sink down into my guts, deep deep down, until the lungs burst, then it wheezes slowly up again.

I put on a shirt and make for the kitchen. From the back of one of the chairs I lift my jacket, a black watch pattern, remnant of a career in education, my badge, my office. The kettle burbles in anticipation of the Rayburn. The tea bags inside have passed the point of infusing and have long since begun to stew. There is a thick smell of tannins in the steam. Sometimes I roll a book up and put it into my jacket pocket. It stops people asking questions.

Today's itinerary. Catriona Calum Alasdair at ten. Conversation in pidgin Gaelic. Tea and cake at twelve. Gossip in English. She will tell me how well I'm doing. Murdo Allan's store at one. Newspapers. Chat with Rebecca, the little blonde girl on the till, a school leaver, full of unintentional charm. Home with newspaper and a tin of soup. Chicken. No. Chicken yesterday. Tomato today. Or oxtail. Let Elgar decide. Tomato, perhaps.

Since it is summer the weather is hot. In southern countries you would take this for a commonplace. Not so here, out on the extremity of Scotland, out at the outer edge of the world's grasp, the dull limits of civilisation. Don't get me wrong. I don't hate this place. Not at all. What I have learned to hate is the way it is occupied by the unimaginative.

The bread in the bread bin is warm and smells stale. It prompts me to put the lid back swiftly without lifting out the bread. I take a cup from the sink and place it on the table top, filling it with thick brown tea. From the fridge I take a lemonade bottle with milk in it. I transfer milk from the cartons to these bottles as soon as it arrives on my table. The Swedes are clever, but they have yet to invent a way of keeping earwigs out of my fridge, and out of the milk. The old fashioned screw top bottle does the job. But I notice a dormant insect amongst the eggs, curled in stasis in the grey cardboard, so I shake it off into the wastebin. This is yet another horror one learns to live with. Funny though. Insects no longer sicken me.

Horny golachs. Na gobhlagan. I don't think I ever saw one in Sweden. It occurs to me I don't know the word for them in that language either. At first when I moved in I devised the most horrible deaths for them. I sickened myself at the seemingly limitless depths of my imagination. Now we share our mutual space. I only have to tolerate the occasional bite on my legs from where they crawl through the bed sheets at night. A doctor friend told me that the bite comes from the mandibles, not the tail. I still like to think they're little scorpions. They look so much alike.

The carpet in the room which overlooks the bay is red like a sacramental wine, a remnant, perhaps, of days when this same building served as manse to the white lime church which stands on the headland above the village, surrounded by a fortified outcrop of black rock. There are stories of its having been a singing school, a college, perhaps – certainly, if the latter be true, the oldest in the country. So much truth was buried with the movement of the dunes or the reconstruction of the faith. And I suppose that most of what we know today is based on academic research into whether or not the Victorians unearthed anything of value other than their own desire and prejudice. No-one has bothered to excavate the fields themselves. I sometimes feel we will never know Scotland in our generation, only what is desired of it, little pertaining to the real past of camped down millions.

Because the church stands a little away from this old manse I am forced, by habit, or by fear of execration, to take refuge in this cell of a Sunday while the beneficent gather like a host around the battlements of rock. On occasion I rise early and make for the northern part of the parish which is sparsely populated and in which, if timed well and if one is canny enough, it is possible to escape incursions into the faith which might leap, guerrilla like, from the lea of the brae. I beat the heathen bracken with my heels. I escape into the open spaces of the past.

Gregor Addison was born in 1966. After attending Newbattle Abbey College, he studied English and Gaelic at Aberdeen University. He has lived in the Western Isles and Sweden, but is now a teacher at James Watt College, Greenock, based in Glasgow.

BRANCHES OFF THE LANGUAGE TREE

SUISAIDH NICNEILL
RODY GORMAN
IAN STEPHEN
KEVIN MACNEIL
ANNE MACLEOD
IAN McDONOUGH
GEORGE GUNN
HARRY GILONIS
ALASTAIR PEEBLES
ALEX CLUNESS
ROI PATTURRSSON
JEAN MORRISET
ANDREW SUKNASKI
SUSAN MUSGRAVE
YUNNA MORITS
GENNADY AYGI
MIKHAIL PRISHVIN
NILS-ASLAK VALKEAPAA
INGER & PAULUS UTSI
KIRSTI PALTTO
RAUNI MAGGA LUKKARI
LAURI OTONKOSKI
MARKUS JÄÄSKELÄINEN
PÂR LAGERKVIST
TOMAS TRANSTRÖMER
EVA STRÖM
TARJEI VESAAS
CHRISTINE DE LUCA
KNUT ØDEGÅRD
ANDREW GREIG
JEFRREY SHOTTS

siusaidh nicneill

IONA, WEST BEACH ON A RAINY DAY

An old left shoe, lost overboard.
An expensive trainer it was once, too.
The knot still tied.
Where's the foot, I wonder?

A frog flattened by the road.
Arms and legs as they last flapped.
An Atlantean survivor
finally tattooed by rubber tread.

Rocks of red and bright pale green
finely etched by sand and weed.
The golden grains and dark, dark wrack
thrown up like a stallion's mane.

A final hard scramble up to Dun I
to reach a well on the roof of the soul.
Heather scratched and fenced in
at every turn –
               I retreat.

from *All My Braided Colours* Scottish Cultural Press (1997)

Rody Gorman

AT THE PEATS

A day at the peats:
It's not the peats themselves
But your company
That day coming into my mind
That keeps me warm sometimes
In the depths of winter.

YOU AND THE SNOW

You penetrate and lie on my mind
Like the snow does at times

On Ben Screel
At the beginning of August.

NEW BOOTS

I put you on like a new pair of boots
And I took off to that mountain
For the first time tonight
And there wasn't a squeak out of them

And neither did they caress my feet
Which remained dry and tight,
Constricted inside them

But maybe all it was
Was that we hadn't become properly acquainted
And that eventually
The heart of them will grow soft against my feet
Throughout the mountain.

DÀ CHÀNAN

Bidh mi cur sìos air duilleig
Mo smaointean ort sa Ghàidhlig
'S gun Ghàidhlig agad fhèin ann –

Nach math gur h-ann mar sin
A bha cùisean riamh againn
'S nach eilear a' leigeil fhaicinn
Nach e cion cànain
As motha tha cur eadrainn
Ach gu bheil sinn dealaichte
'S nach tuig sinn cainnt a chèile.

TWO LANGUAGES

I put down on the page
My thoughts about you in Gaelic
And there's you with no Gaelic –

It's just as well
That that's how things were with us
And that it doesn't show
That it wasn't primarily lack of language
That came between us
But that we're divided
And we don't understand each other's speech.

Ian Stephen

ON *AN SULAIRE* – IONA TO CARSAIG, 26TH MAY 1997

a brutal bursting
disruption of lifetime's fetch
a bitter straggler

    *

our bow's a chisel
wet aftermath of seriphs
a hissing exhale

    *

helm hard till we luff
three reefs make us comfortable
fisherman bites bay

Kevin MacNeil

From the Gaelic & the Japanese

HAIKU

In theekineeng mist
a forin joke sails beetween
the hills and this boat.

The lash of a harsh
Loo-is rain. Endless moo-ur
and God's loneleeness.

In the graveyard a
ministur playeeng frisbee
with his eldur son.

After the Gaelic of Iain Crichton Smith

A morning rose
dawn sunlight
– just so, your breasts

Summer in Lewis (after Buson (1716-1783))

on the tshurtsh bell
purtshd sleepeeng
a butterfly

Anne MacLeod

ANOTHER WEDDING

I sit on three-week grass, grown soft and long,
to view the photographs; another wedding
another clear blue sky, another world
and girls, big-boned and blonde, with Lewis eyes.
They did you proud in Canada. (The bride
forgot her veil. The dress looked fine – but think,
three bridesmaids and no veil!) That white dress swirls
from Sarnia to Callanish, the eyes
from Callanish to Sarnia, unveiling generations:
sisters, brides, spin gently in the clear
unyeilding years.

Your white dress swirls around you, and below
the lingering island waves a calm farewell
in moonlight and the psalm that single ray illumines;
this bride forgot her veil. Your camera
exalts her open beauty, while the eyes
from Callanish to Canada seal disparate generations.
there is one more distant, diffident; we see
you
    dancing
through the trees.

from *Standing By Thistles*, Scottish Cultural Press, (1997)

Ian Stephen

FOR WILLIAM MACGILLIVRAY, NATURALIST AND ARTIST, (1796-1852)

*the half-checked purlins
on Scots larch trusses –
vernacular Garenin*

from Garenin

(wolf spider)
    treading close the turf
    from Garenin to Dalmore

    wary of the wolves
    which carry, at speed,
    their pregnant nests

    the underswell
    at the linked stack

    through the strata
    of the tinted interpretation –
    MacGillivray's shoal

    of Atlantic wrasse;
    the dragonet
    and Peter's dory

On Toddun

*(cuckoo, merlin
golden plover)*
    cuckoo in distress
    in the gradual scrub

    merlin cuts the face
    of vole-grey rock

    in gold deer-grass
    a parent plover

Pabay Mor 1

(velvet crab, heron)

a cistern in drystone
wetted by flood

now holds velvets

greens for the greys
of crabbing heron

Pabay Mor 2

(shag)

whites from the nests
mark a rounding

open to Bernera
the arch resounds

sea-hungry bass;
aerial treble

cresting for
sandeel sheen

George Gunn

A KYLE ON THE NORTH COAST OF SUTHERLAND (after a photograph)

It's summer, a woman & a man
sit on a wall in front of a
silver beach blue sea
she is about sixty, maybe younger
he is nineteen, maybe older
there is a small dog, a Jack Russell
it's summer, it looks like
Bettyhill, the woman seems as if
she is waiting, the man has
his left hand on his neck
& a pint, three quarters drunk
in his right hand & for some reason
everything has Sunday written on it
the drystane dyke behind them
is a perfect liquid skin & a
hidden sign asks & hopes that
everybody does thirty, some whins
grow, sheep graze, it's not
an unusual picture except
that the woman would duly kill
herself, the dog would die of cancer
& the man would ruin everything he touched
the sky is thin blue like calor gas
there is the suggestion of blue mountains
somewhere far away & it can be said
that if it were not for the exactness of this photograph
it would be about anybody anywhere
anyway it's summer

Ian McDonough

A WALK ACROSS THE CLEARED AREAS

Altanduin
    The ruined land
    is whispering its history
    to switchback streams
        boundless sky.

Tuarie
    By Strath na Frithe
    the past is darkening.
    Pockets of sun
        illuminate
    a bare-stoned future.

Achrintle
    Where children sang
    wraith mists
    an empty moon.

Badanraffan
    Sunk in ferns
    shadow homesteads
        indistinct
    something not-quite-remembered.

Ascoile
    The path dips
    into activity,
    worked land.
    Looking back,
    a sign – 'Danger – Stalking in Progress'.

Harry Gilonis

from READING HÖLDERLIN ON HOY

On the Howes of Quoyawa, on
the Knap of Trowieglen, fish-
bone-widths of snow
silver in the sunlight:

to move, pathless, among rocks
and heather, aflame with
quiet fire; to trace
the course of streams,

to learn to tell
white hare
from white boulder,
the specific

names like morning breezes
and their absence, too, cooling
as the un-named lochan's
slaty water...

On the crags
by the burn of the Kame,
the bare stones
of language:

under that dark light
no word
like 'flower'
will flower...

here, twin slabs of rock
and it clear that, once,
a rocking-
stone sat,

poised, local,
wavering
between the total
and the particular,

set wobbling
by the weight
of the immanent
moment,

by thought
rocked
as if settled
on water:

on a mountain
on a slope
on a hill

adrift
in un-
certain sea

without ebbtide
without oar
without rudder

Rackwick, March 1995

Alistair Peebles

THE ISLAND'S FULL OF HOLES

She said. The caves go miles. The bedding planes
Will slide us, from the Sneck to Hametoun,
Off to Spain: Agricola's olives, vines.

Imagine Foula, north west, fathoms old
And whole. A field of oil-soaked rock, drilled down
And pressed for joy of engines, plastic, gold.

They capped the Lum of Liorafield, and sealed
It with a curse. Put your ear to the ground,
To the rumble of stone, the endless fall.

SHEEP ISLANDS

Flying over Shapinsay's ten-acre fields:
It's too sudden, this closeness, these shipshape roads.
I've come from a sheep island marked out in folds,

Squared only by map lines, fenced in straggles
Round the heel-backed hills. Nine days handspun, loosed
From the check-in, now I'm back with a bump,
Down here. Moments since, up out of Sumburgh,
Foula lay blue and alone. I gasped, but missed
Her going through clouds: stared on till I saw this.

AT RUN HOEVDI

We were strolling through the clifftop bruck.
She frowned. In Volta, Niger, Mali, Chad,
It was the same: old cars being left to rot.

I thought of Edwin Muir's survivor, glad
To let those dank sea-monsters turn to mould,
Of Oceanics crumbling on the Shaalds.

I said, Look round, they'll wear down soon enough.
But she knew what was dying below the surf.
Of rust: the deserts, what was left of us.

from *NINE DAYS ON FOULA* (1996)

Alex Cluness

MØRKADALUR, FØROYAR

1 – FØROYAR

A strange bridge,
stepping-stones back
to a different time.

An island mist
that rises black
like instinct

from sudden
white.

2 – MJØRKADALUR

When I first went to
Faroe as a boy,
a drive from Tórshavn
found a valley
filled with mist.

My father, who'd been there
many times, said –
This place is always
full of mist.

And I can't remember
if he told me then
that it was called
'The Valley of Mist',

but over the years
I have convinced myself
he did. It gives me comfort
(I suppose) to imagine a place
that does not change,

where no one sees the things you do.

3 – THE VALLEY OF MIST

I have seen the NATO base
cut into Mjørkadalur,
but don't remember it.

I picture a turf roof
like the furrowed brow
of the Hotel Borg.

A glint,
perhaps binoculars,
no more.

Because it is not
in my memory,
it belongs in books.

(I have seen it,
as I say,
but it has gone)

I remember only mist,
a car descending,
my father's hands on the wheel,

laughter in the back
and all around me
mist-strange glimpses.

Green breaking
through the surface
like trench gas.

And years later
this sudden sense
that here was a place

for keeping instinct,
a place that does not change,
where no one sees what you do.

4 – POSTCARD

Mjørkadalur's real,
but the Valley of Mist is not –

it is a creation only
in imagined landscape,

carved by need
out of time.

(It exists in my memory
because I want it to)

5 – A FAROEMAN, LIKE THE WORLD

Under grass eyebrows -
a subterranean Saxaford -
the glass eyes of NATO
glower over Mjørkadalur.

While up in space
a satellite can shift
to see a man in 'Havn
roll a cigarette.

And when I think
of him, I realise –
that everywhere changes,
that everything's seen.

From the Faroese

Roi Patursson

DRUNK

Curiosity and thought
were thrust into the deep
by silence and you all.

I went
like a caged beast
along a stony, high wall
and searched.
I knocked uselessly
till my will and hand were broken,
I wanted out
but where was the hidden way?

I found it,
it was a small lie
that God as well as all good men were in on.

It brought the planets
and endless space
to life,
it quickened slowly,
no flame
just embers
a red glow from the rest.
In the distant hum
from the planets
you all vanished
an overbig nothingness
was born again as music,
sounded for a moment
and faded away
to a man
who mournfully sat prisoner
of silence and you all.

But the hidden way
lures into space
where longing just grows
because freedom is unknown.

trans. George Johnston

From the French

Jean Morriset

geographies geographies

*for jack kérouac, ernest hébert, paul théroux,
david plante, clarck blaise and ... all the
'franco'-writers of invisible america!*

geographies geographies
me ajudem por favor

nomadic snows on the icebanks of the open world
spellbound memories of drifting snow
prayers the old sagamos engraved
come to my aid

tell me
the lives of all my ancestors

tell me
all the dreams of those who disappeared
in the current of the americas

racontez-moi l'histoire de ma vie antérieure

i am part of a black out
all trace of its memory gone

i am part of the memory of some other memory

geographies geographies
me ajudem por favor

i am of the tribe
that covered ten thousand years of ice-age
in the wandering of the species

i am of the tribe
of the overflowing rivers' limbs

lakes and peats are the mantle of my people
endlessly wandering the fringes of a future
in everlasting portage

i am of a people from whom was cut the mississippi

*

memoria oh memoria
memoria do fogo
memoria do gelo

memoria da memoria

rime-burns on the tongue
the banked-up earth's white sweat

sunset strokes on the nape of time
twilights spilt in the plough-wake of history

memory of some other memory
memory of the sleeping night

meditation of grass under snow
memory of the vertebral source

**memory of otters coming each spring to lie
on the breast of the sandbanks in the strand of your loins**

oblivion-memory
me ajude

i am of a tribe
history left in set-aside

i am of the butterfly clan
of no fixed place sucking from tongue to tongue
among the juices of the transhumance

memoria oh memoria

fevered memory fallow memory
fusion-memory erosion-memory
me ajudem

i am of a tribe
of pollens glistening with the moistness of being

the wind keeps the archives of my tribe
lichens encrust its mythologies

rolling carpet land
criss-crossed by frontiers
never belonging to it
beneath the wave of decrees-in-council
my people has seen its geography
give way beneath their moccasins
to relieve them of their wandering

all my people's wild lilies-of-the-valley were uprooted
to relieve them of their liberty

*

memoria oh memoria

oral memory
voiceless memory
illiterate memory

memory buried beneath the mangroves of the great north
memory of the sleeping womb below the cogwheels of night

oval memory
migrant memory
mist-flake memory

memory of silence splashing on the crystals of the sun
memory of the quaking ice

memory of womb-streams
on the western buttress

memory of the hares leaping
in the undergrowth of dream

memory of the grand coulee crossing
slipping down to the drowsy beach of oblivion

memory of the unspoken things of summer

tell me tell me
all that they were forever forgetting in order to survive
if you wish that i too forget
in good patriotic faith

men came dressed in red they burned our houses and our schooners
they pillaged our fields and mowed down our dreams all the river
long and so we decided to burn our memories so as to dowse their
fires and undo all trace of their passage
men came dressed in red they wanted to carry off your mane of hair
and tear out your bracken but you escaped like a flight of teal above
the ebbtide in a joyful clamour of wings
a great boreal frisson crossed the twilight here and there and we
reappeared on the far side of winter like white partridge with purple
petal-feathers that no minus-forty would be able to suppress
**and so it came about that we danced like snowhoppers that no
glaciations could eliminate**

memory of all the unspoken things of winter

tell me tell me
all that they were forever hiding in order to survive
without realising it was killing them already

memory of the footsteps walking
in the kitchens of all those little canadas nice and wasp
behind the window of assimilation

those who ranged america from top to bottom long before the yankees arrived had only one haunting fear to abandon their original archaeology to forget their history by the very cult of the past in order to become *bang-up-to-the-minute-americans* – just like the greeks italians portugese or chinese – and gain at last the pleasure of the same democratic segregation as all the rest

'Tell me, tell me, don't you remember? Speak to me that language you used to murmur when i was a kid. It was so soft. Fresh as a pillowslip. Oh, sing me again that chanson you used to sing so quietly in the twilight rocking-chair. That lullaby. Lul-la-by! That libellule. Li-Bell-Yule. You no longer know it? You've forgotten the word libellule again. I can't believe it. It took you a full month to extract it from the basement of your fading memory and you've lost it again. I just can't... Your name is the last drop of french i got. Nothing else. Why did you quit speaking my language? You should have quit smoking instead. Why did you quit dreaming our own memory? Why? Why? Why? Why have we so meekly allowed our own memory to be wiped out?

Libellule. **Libellule. Li-Bell-Youle.** Oh, what a splendid lullaby we were?'

*

memoria memoria

memory of other languages
i spoke once
whose every curlicue i have fogotten

memory of another tongue
vanished beneath the water line

huron words
iroquois words
muskeg words
words gone in an open boat
on the geography of the tundra
never to return

nee-zee t'sontsi-tzè
bout-tchi-à-tcho
maçi tcho

ice words moistening the lips
of the raging river
eskimo songs dancing
on the northern lights of february

*

dis-moi mon frère
in what language did our ancestors dream
on the tide in the depths of the woods
in what tongue did our ancestors trace
the trail to be opened after the storm

fala para me *irmãzinha*
in what tongue did the old ones blether
to the racoons in heat and the screwing foxes
in what tongue did the sorcerers of the isle of orléans dream
just before they turned themselves into wendigos

mech' mech' mech' don'
marche marche marche donc
growled the old eskimo couchines of koyoukon
to their huskies

ravages chicoutées michipichou
någane barcanes bouscueils

where do they arise from all these words
fleeing academic control

ouapiti chikok carcajou
aglou oumiak eekalou

but where do they come from
all these sounds

*

memory of the loss of a river
memory of a geographic re-emergence

little sea-faces
glaciers minstrels
forests porches

water-lily memory
marrow memory

fugitive-slave memory
memory of the skin

a pele como memoria
o sotaque como pele

skin becomes accent
memory becomes colour

*

memory-tide
come nearer still
take my river by the hand while i tell you

sitting in front of the patriarch's shack
at the turn of the bay-of-shikoks
one day i saw the city of chicago arriving by train
towards the end of an afternoon of 'scattered showers'

i was always there
i was always present
behind the history of this continent

at the corner of 51st and 49th
during the night of the long knives
juste derrière la clôture

j'étais là

tu ne m'as pas vu dans l'ombre
keeping an eye on a cloud of mosquitoes

j'étais toujours là
et tu ne l'as jamais su

j'étais là
debout sous le soleil
when the missouri bore her last steampacket

i was there
at Saint-Louis-du-Mississippi
when the corsair Jean Lafitte died like a prince
telling one last loving tale of a ship-boarding
to warm up an old mestizzo chilled with the fever

j'étais toujours là

memory frightened before its own recollections
why have you pretended not to see me

i was here there everywhere
and
while they were trying to tear out my witnessing eyes
i was whistling in their faces
the song of wild oats and wild rice
and escaping from them once more
in a maelstrom of two-step rigadoons

*

sight of all the caravels
swept in by the solstice

geographies geographies
névés névés névés
me ajudem

see how june's great tide
has rolled in her words right to the northern cross
and bathed her rushes as far as the southern star

mirage memory
rose-filter memory
memory of the future

may the mist's clarity come
and the smell of the fresh light come
into the thighs of dawn

i have crossed all the rapids
and come down every isthmus

i have walked through all the forests
and crossed all the clearings

illiterate writer of the geographical crossing
i want to swim in the debacle of the centuries
and retake the vanished space

i am the fugitive son of a more than human race
one day you will hear tell of me

trans. Tony McManus

Andrew Suknaski

SĂT

'om vinit din šat wood mountain post.
'"i came from the village wood mountain post"'
they would say to someone enquiring
or 'om vinit din šat moose jaw'
they remembered as they left the prairie
their creaking wagons and whinnying horses
resounding through the hills
where they were told they would find their homesteads
and a new home

'om vinit din šat ...' murmured the faint ancestral faces
peopling the memory of the old home
these men and women knew

with wagons and horses
they carried their supplies south
from moose jaw
the journey taking five days
till they arrived on the northwest side
or twelve mile lake
it was may and growing warmer
while they camped
and explored the county for four days
till they all agreed
on what would be known as šat
'village... or settlement'

they built sod houses with brush and sod roofs
and lived together for two years
till surrounding land was surveyed
and became available for homesteading

there were seven families that year of 1906
and their names were:
yordachi adamache
costantine mehiau
john stefan
dragu cojocar
nita cojocar
john stoian
georg chiro
badar tonita

2

vasile tonita remembering what his parents told him
'the first few years
the rcmp checked the people once a month
to see that all was well
and that there was enough food and fuel...
the police from the old post gave a permit to šat for $2
to cut some wood up in the wood mountain hills
it was said that wherever they unloaded the wood
that was where they would get their homestead
but they dumped it all in one place at šat
so it didn't quite work that way...
before they all homesteaded
the government gave them authority to break some land
they sowed about 20 acres of oats
they bought a ripper somewhere
and cut the oats and raked it into piles...
the fall of 1907 at harvest time
the men worked on the threshing crews aound regina and
    dysart
pitched bundles for $1 a day
a day and a half of work meant 100 pounds of flour
anyway they made enough money to buy supplies in moose
    jaw
for a second winter'

mrs. vasile tonita remembering what her parents
the adamaches told her
'my mom and dad went to dysart with a team and wagon
in the spring of 1907
they led back a cow and calf...
they also brought four or five chickens
when they were coming home
they say everyone was there at the edge of šat
they were all so happy
they had cheese butter and eggs for the rest of their stay
in the winter they played cards a lot
visited one another and told stories
about the old country'

vasile's final words
'and i guess it was lonely sometimes
waiting... you know
waiting for land of their own...'

3

new year's day 1977 and two miles beyond flintoft
riding in the warm truck with lee and marie soparlo
marie the daughter of vasile tonita
a silence broken when lee points across bright yellow stubble
above glaring white snow
'you see that single tree just beyond
mind
where the snowfence has fallen...
just to the right of the tree
and by the crick in that bit of coulee
down from the railway track... that's
where šat used to be...

from *Wood Mountain Poems* (1977)

Susan Musgrave

SKOOKUMCHUK

I guess it's in
my blood
to want to be like
Emily Carr. I don't know
much about her
but we've been to
some of the same
places.

The north is
the end for me –
I'm in love with a
man I'll never
meet.
Indian Jimmy from
Nanootkish (was there ever
such a place?)

Emily and I
shared him for a while –
I know that. He was
impossible to paint
and what's more
she found the forest
a deeper attraction.

The eagle
eats the land,
I write in my
journal. A nurse
wraps my wrists and
says next time
don't use
third-rate machinery.

*Give me back
my own;*
I want to go where
Ninstints found his
name.
Jimmy rows
further into the
sea-drift
Emily says
it's too rough
to go sailing.

She paints
the unexposed skin
the masks behind
loss. My notebooks
have been empty
up until now.
I write often to
Nanootkish
but my letters
always
come back.

from *Selected Strawberries & other poems*, (1977)

Three Haida Songs

LOVESONG

Beautiful is she, this woman,
As the mountain flower;
But cold, cold, is she,
Like the snowbank
Behind which it blooms.

THE BEAR'S SONG

I have taken the woman of beauty for my wife;
I have taken her from her friends.
I hope her kinsmen will not come
and take her away from me. I will be kind to her.

Berries, berries, will I give her from the hill
and roots from the ground.
I will do anything to please her.
For her I made this song and sing it.

CHARMSONG FOR FAIR WEATHER

Good sun, look down upon us.
Shine, shine on us, sun,
gather up the clouds, wet, black
under your arms
so the rains cease to fall.

Because your friends are all here on the beach,
ready to go fishing – ready for the hunt –
So look kindly upon us, good sun!
Give us peace in our tribe and with all our enemies.

Again, once again we call–
Hear us, hear us, good sun!

trans. C. L. Skinner

Yunna Moritz

From the Russian

BAD WEATHER IN DICKSON BAY

For two weeks the lightning-shattered sky
Has unloosed a flood onto this piece of land.
In the harbour blue from the cold
The ship's languish in a manly way.
The sea has drawn a dove-grey
Soaking mist over the shore.
Here it is impossible to invent things, to deceive,
Hope being so precious.
By sharp, tormented masts
The faded flag is thrown against the sky.
The five fingers of the hardy island
Sometimes clench into a fist.
And then, in the endless nights,
With undisguised sadness and pain,
The human-voiced sirens
Break out like trumpets!
The salt water writhes,
And fires tremble on the island.
The ships are feeling bad, feeling bad...
They are bursting to put out to sea!
Two weeks – I can scarcely believe it.
I live on a frozen shore
But in some secluded corner of the heart
I always keep hope.
Wait for me, do not feel resentful,
Write to me that the woods are in blossom.
You understand, I must first see
The ships leaving here.

trans. Richard McKane

Yunna Morits

TO YU. V

I am going away, going away
Into the bright, distant steppes,
Where rushing dogs follow the taut
Tracks, woven into a chain.

There birds of prey drift,
And the sun burns down mercilessly,
And catching the scent of dogs,
Hares' hearts burst.

I will lay fires, and cook
Supper for the hunters,
And will forever forget
How I need to talk with you.

Exhausted I shall fall,
Yet mark the howling of the wind.
May sky, mountain and gorge
Guard me against you.

I will begin to sense the wild beasts
In the fiercely trampled grass.
One day, these brave folk will
Confide their trade to me, –

And the moist body of a vulture
Will hang from a strap at the waist.
I am going away! What do I care
How you live without me.

trans. Richard McKane

From the Russian

Gennady Aygi

BURNING – DURING HARVEST

to L. R.

I bury myself – in the harvest: oh
this fire – that once tempered
our fathers' patience – like the guiltlessness of earth! – and the
                                            Land in simplicity gleamed
as if – it was ringing: and into its heavens ascended
meaningnames
of Poor Things – and translucence of their goodpresence
forged the heart of one who inwardly leaving for other walks
knew – of non-return
to this hearth .... – long ago ceased ringing in
                                              anonymous common-space
that closing of the circle – in absence of light-foundation:
            of the fathers' shining as they sang with shoulders
as with faces of sky and soil! – and today
only the silence of the Word
gleams in the facelessness of the world .... – yet it was
             simple: yes even the likeness of timid whispering
was – like this heaven-face – everywhere deep
and dear – without limit

1983

trans. Peter France

Gennady Aygi

FORESTS – BACKWARDS

in the mist
of the shining of home
have remained like islands like pearls
the forests
I never reached –

– I remember something of childhood: little shoulders
          visible in them – white inclining to the fields
or else suddenly
complaining – movement slackened: more in sorrow
than the visible – there on the inaccessible
edge of the forest –

(there were some – I saw from so near
but soon
there was only the wind –

easily – as if in the wind – I learned
easily to understand that there is no return) –

in the light of valley-crossings
it seemed – that children awoke amid grasses
and their singing sought words – somewhere nearby
as if from there
it seemed –

in the mist of the shining of the world
they remained like pearls like islands:

more painfully than in life – to shine

1985

trans. Peter France

AND: THE PLACE – OF THE FORMER SIGN

and that which once
was glimpsed – as if in a vision:

– in sunset light
beauty of a bowed head –

(from goodness created or creating goodness) –

remains – to shine:

there – where once the people
was – a field and the place:

of head-radiance:

– centre of worldspace! –

(transparent with souls – as if with one:

special – wind) –

oh – persisting:

(there has long been for no-one even the breath of the gift –
to see and remember) –

in the world (even if abolished) amidst Field-Russia

beauty of a bowed head

1985

trans. Peter France

Mikhail Prishvin

From the Russian

from PHACELIA

THE FORK

Three roads lead from the signpost; in different ways each is catastrophic, for at the end lies the same destruction. I am not going where the road branches, but coming back from the other direction, towards the post where disasters converge. Gladdened by the post, I return homewards along the one true road, recalling my troubles at the fork.

MY DEAR OLD SAMOVAR

Sometimes there is silence, such clarity in your soul. You look at folk so carefully, and if a person seems beautiful you admire them, and if a person seems ugly, you pity them. And in every single thing you feel the soul of its creator.

Here I am, lighting my samovar, that has served me thirty years. And I take care that my dear old samovar doesn't drop a single tear when it comes to boil.

Joy

Sadness, building up inside your heart might one fine day spontaneously combust like a haystack, and everything will burn in fire of extraordinary joy.

Nils-Aslak Valkeapaa

From the Sami

My home
is in my heart
and it moves with me

In my home where the *joik* lives
there is the laughter of children.
The bells clang
the dogs bark
the lasso whistles.

My home is in my heart
and it moves with me.

What will I say brother
what will I say sister?

They come
and ask where I belong.
They have documents with them
they say
this belongs to no-one
this is the State's land
everything is the State's.
They search in their thick, dog-eared books
they say
this is the law
and it affects you too.

What will I say sister
what will I say brother?

For they ask
where I belong

I hear it
when I close my eyes
hear it
I hear
a place deep inside me
I hear the earth boom

thousands of clovers beat together
I hear the flock of reindeer thunder
or is it the shaman's drum
and the rock of sacrifice?
I sense
that somewhere inside my breast
it whispers talks calls cries
the clamour fills
my chest from wall to wall.

And I hear it
even though I open my eyes
I hear

Somewhere deep inside me
I hear
a voice calling
and hear the *joik* of the blood
deep
from life's boundary
to life's boundary.

All this is my home
these fjords rivers lakes
the frost the sunlight the storm
the night and daytime of these moorlands
joy and sorrow
sister and brother.
All this is my home
and I carry it in my heart.

trans. Kenneth Steven

Inger and Paulus Utsi

From the Sami

NOMAD TRAILS

The old one sits
in her spring camp
longs for the summer country
beyond the mountains

Follows paths, pats the reindeer bull
sheds tears
wades streams, unties the knot
stuffs dry grass into her shoes

Thinking – they will disappear
the tracks left by the Sami
as the Sami of olden times foretold

trans. Kenneth Steven

Rauni Magga Lukkari

From the Sami

When you told him he came
you made coffee
laid out food
as if he couldn't manage that
grown man
but for you
he is a child
whom you wash
clothe
feed
and put to bed
except
when he is drunk
then he is the master of the house
big
and you are afraid

trans. Kenneth Steven

From the Sami
Kirsti Paltto

So small and lovely you are,
my boy,
but one day you'll be big and strong,
my son.

Now I feed you,
my boy,
but in the end you'll take on your own shoulders
my people's inheritance, which I will leave you,
my son.

I come from an age
which smashed and destroyed
broke in pieces
carried many off

stamped out the embers
which I should have warmed myself over

trans. Kenneth Steven

Lauri Otonkoski *From the Finnish*

ON THE EAR'S WALK

The landscape's deepest melody flowed on
    over the banks of the resounding Middle Ages.

Do you hear, do you hear it
the way a snail hears,
that snail there who teaches,
learns from the earth's replies, learning
the snail hears and gets there,
gets there for sure
even the slow one gets there,
even the slower one will
then get there, it will
surely get there, into the pot.

trans. Anselm Hollo

AROUND ZERO O'CLOCK

Go ahead, be the shape of an angel, be, be
    be, be a screeching
  hatful of sleepless night   it dresses
even the seagulls in diver's suits, be
  be lazy intellect and come
to bed
be manager of nightmare
  and conqueror of desire

                to say

Be the disease of saying   Be the lifelong remedy
  which   whether you take it or not
    certainly kills

Be the one who no longer is
    a dab of freedom of the void, a flight of three strides
out of thought's night   be

Because I'm fading

trans. Anselm Hollo

Markus Jääskeläinen

From the Finnish

The smell of candy stuck to my clothes and came inside.
I washed my face, soaped my hands, it didn't go away.

This is a candy planet,
my friend said. All you see
and hear is sweet. The people whose hands you shake
are sticky. You can't get rid of them, nor they, of you.

I took a deep breath.

We like the smell of candy, she said,
it is the smell of home.

This is not my home, I said.

But you live among us, my friend said.
You are alone.

You had a dream. In the dream, we moved.
The house was new but the things were old, familiar.
We were home when there was a knock on the door.
We didn't open, even though they started rattling the door.
But then they came in, these men,
looking for the belongings of some dead person.
You said, there's no one dead here,
you're in the wrong house.
This is the house where love lives.

trans. Anselm Hollo

They lie in the flurrying snow, languid as a naked woman taking
 a shower,
the mountains, their luscious thighs ajar; under snow-white skin,
confident rib-tongues curve down the gully
where a lone skier slides and struggles in unbroken snow

A dense stand of spruce grows from her thighs, moonlight
shimmers on her flank, her hair is green

A hundred miles long, face hidden under the covers, out of the
 smoke
droplets emerge

slow is her breath in the wind, waiting for spring, under the snow

No one can conquer that vision, move it, bury it,
stitch it shut

she has come without being invited, living rooms grow inside her,
mice rub their whiskers in her hiding places,
obedient, the sun sets behind her, opens the dark door

trans. Anselm Hollo

Thomas Tranströmer

From the Swedish

The sun is low now.
Our shadows are giants.
Soon all will be shadow.

*

The night flows westwards
horizon to horizon
all at the moon's speed.

*

Oak-trees and the moon.
Light. Silent constellations.
And the cold ocean.

trans. Robin Fulton

Eva Ström

From the Swedish

THE RADIANT EARS OF CORN

After hymns and sermon the book-pyre was lit.
The uprooted bramble broke out of the grave.

The pages burned. Soot flakes fluttered on the wind.
The murdered women all had first-names.

It was in the crèche the rape happened.
The mysterious person took shape and started singing.

She had a thread in her hand. She had met monsters.
The True Congregation was reading The Acts of the Apostles.

The burnt books were valued at 50,000 silver drachmas.
The evil spirit had maltreated them naked and bloody.

The parachutes were released. The orthodox found themselves on
                                              the pages of newspapers.
There was no longer any market for small silver temples of Artemis.

The one road darkened and coiled its way out of history.
A new cross-breed grain took shape. In amazement the people
                                              watched the radiant ears of corn.

trans. Robin Fulton

Eva Ström

EZEKIEL

Can you hear the singing from inside the church
Who is singing? Cracked voices, old women?
Is someone lying in the coffin? Has it been emptied, plundered?
Is there someone there, dead bones, clothed in sinews,
dead sinews, grown together with muscles, flesh?

In the sand-box a bird with bloody feathers
is covered with tiny flies.
The bacterial process continues.

The hymn falls silent, the stone bleeds.
The prayers chafe against the stone. The voices chafe
against the prayer. The hymn chafes against the stone.

trans. Robin Fulton

Pär Lagerkvist

From the Swedish

I listen to the wind that sweeps away my prints.
The wind that remembers nothing
and that barely understands or cares a jot about what it does,
but which is so lovely to listen to.
The soft wind,
soft as oblivion.

When the new morning breaks
I shall wander further.
In the windless dawn I shall begin my wandering afresh
with my very first step
in the miracle of unstirred sand.

trans. Anthony Barnett

From the Danish

Werner Aspenström

THE STRANGER

Brotherhood, jostled together in buses on twisting roads
or on ferries in a gathering storm
Sisterhood, the dread of cancer in waiting rooms
with art-club prints and yellow creepers.
What an unpleasant sort – him, over there,
no need of consolation or illusions.
The shivering pigeons in the square outside
neither touch nor disturb him.
Twice I've asked him who he is
without getting an answer.
You'd think he belonged to some other branch of the language-tree
or was only a shadow
with the indolence typical of shadows.

trans. Robin Fulton

From the Norwegian

Tarjei Vesaas

SNOW AND SPRUCE WOODS

Talk of home comfort –
snow and spruce woods
are comfort.

From the very first
it is ours.
Before anyone has said it,
that there *is* snow and spruce woods,
it has a home in us –
and then it is there
forever and forever.

Yard deep drifts
around dark trees
– that's for us!
Infused with our very being.
Forever and forever,
even if no one sees it,
snow and spruce woods are ours.

Only a snow-covered slope,
and tree upon tree
as far as the eye can see,
wherever we are
we turn toward it.

And carry within us a promise
to come home.
Come home,
walk across,
bend branches,
– and feel as it courses through you
what it means to be where you belong.

Forever and forever,
until our inland hearts
are stilled.

trans. Anthony Barnett

Tarjei Vesaas

THE BOAT ASHORE

Your still boat
harbours no name.
Your still boat
has no harbour.
Your hidden boat ashore.

For this is no harbour –
The leaves shimmer on spring nights
above the readied waiting boat,
and strew their wet and yellow
onto its thwarts in October,
and no one has been here.

But here there is a pull from endless
plains of sea,
where suns rise out of the depths
and the wind blows toward the harbour beyond.

But this is no harbour either,
only a place with the pull and call
of still wider plains,
heavier storms along the coast,
and a larger boat in the evening.

Your still boat
settles slowly.
Your hidden boat ashore.

trans. Anthony Barnett

From the Norwegian
Sigmund Mjelve

The dreamer and his dream
were there already
in all the known rooms
and all the unknown,
in corridors with all their doors,
galleries with paintings and processions,
in halls where all was still
and archways in courtyards.
The light came and went,
rain and night.

But the dreamer and his dream
went out into the fields,
vanished in wind and light,
went far away.

Here I stand, exactly
here on two legs
between a certain

birth, a certain
death I brandish
my word torches,

I conjure all
worlds to greatness,

to love, peace for all
nights and days.

trans. George Johnston

Sigmund Mjelve

We see the light in front of us
before
we see the light behind.

It blurs our eyes
and becomes
light within light.

But behind us
light too
is light,
it makes our shadow.

So, in a poem
Li Po says:
Tonight

With the moon at my back
I dance
with my shadow.

Horsemen ride
home at evening,
ride slow.

Their bodies bear
their day and their morrow,
the blue air,
their horses are
heavy as they.

Like smoke
from a smouldering fire
rises the western sky.

trans. George Johnston

In the Shetlandic

Christine de Luca

MYL-GRUEL AT DA MILLENNIUM

In Norrawa we had a dish o rømegrut
to mark midsimmer: sweet mylk an bleddick
cookit tae a gruel. You supped again
your boyhood: aye myl-gruel in Vidlin

at johnsmas time: a treed spun
owre a thoosand simmer seas, at we,
wan generation, hed riven in favour
o low fat fromage frais.

Your telt o johnsmas fires: emmers
o Viking taands at burned
on ness an taing, an still burn ta dis day
in Norrawa, atween fiords.

You minded stories tö o Dutchies
an der johnsmas foy in Lerwick,
da hidmost day afore dey set der sail
ta follow shoals o simmer herrin;

an hearin o da gig at cam at johnsmas
fur da gutters. Bi da saison's end
dey'd hae penga i der peenies, a string
o jet black beads, a laad, a haert-stane.

Neest johnsmas we'll mak fine gruel
wi crème fraîche fae Isigny.
Wi a spön o honey hit'll mak a feast
fur John da Baptist's day.

Wir peerie foy'll see nae Dutchies
dirlin doon da street; nae gig
o gaffin lasses, arles lang ta'en
afore da aggle o da creel.

We'll no let johnsmas come an geng
ithoot a bonfire, an wir myl-gruel.
We'll celebrate da langest day, an trivvel
till we fin dat tinnest treed ta tie secure.

From the Norwegian
Knut Ødegård

OCTOBER, ORKNEY

I (Oxen)

Great gray oxen in thin light,
towards evening stand and breathe, without moving.

Sudden dark, foam
in white sea comes towering in
towards them, in the night. Beginning to budge, oxen,

towards byre and house. For oxen feel the cold too
in October dark, oxen too see the moon
whet her sickle over our tiny islands
in boundless, dark ocean.

II (The Cathedral of St. Magnus)

Gray oxen walk over frozen earth.
Stop for a moment, stare at St. Magnus' Cathedral.
Oxen think not: Cathedral, see! like a
whale stranded on a stormy night.
Pull at the grass and plod onwards. Cold,
they long for home.

St. Magnus' Cathedral like a whale stranded on our rocks
from the mighty deep! Dreams therefrom
move fish-tail slowly about
as vague figures among drowned cities, myths, wrecks:
Jonah! cries the priest, triumphant, and calls in
to mass, of flesh and blood, into the whale's belly.
This dark river of blood is endless, this flesh
shall never, never be consumed.

Oxen snort and shuffle about all night.
Think nothing of Earth turning round
and round endlessly, whales in its deeps
and oxen that snort and trudge their way home
on a cold October evening on tiny islands in the sea.

trans. George Johnston

III (George Mackay Brown)

Taxied one evening from Kirkwall
to the poet in Stromness, George,
with a whisky bottle gurgling in the back seat.
Left hand driving there, and miles
not kilometers showed on the speedometer,
vibrating on the dashboard: the driver grinned

like a whale, but kept himself safe on land.
At last the tyres shrieked high among the cobbles and crooked
streets of Stromness. The poet was awake and I
pulled the cork: 'I sit and call folk up
from the grave', said he, 'from the quiet churchyard:
Andrew, I call, who was married
to eight dry women and buried seven of them and then
had his eyes closed at last by cold
girl's fingers. Jeems, I call you, you
who flew over our five islands and fiddled dead drunk
with your bow over every last heart for weddings,
burials and births. And on Winterbride
I call, who put on her black shawl
when they came and said, 'A wave took your beloved,
Jock' and forth she went with a fish-knife; Lay and washed
in the same waves, the black bride flung herself from cliff-side
into the white waves at night. Did she
strike knife and its fish-guts into her heart as she fell? I
call on you, cry you out of your graves!'
said George. 'And on all you with Biblefingers
and fish blood in your beards under stones here
in Stromness!'

Bottle empty now, and the driver looked at me
with his underwater grin, looked at his watch
that ticked and went: 'Ticking and going,' said he.
'Time.' Out in the night, wing-flapping, crying and shrieking
in the air, and down in Scapa Flow cruised a great whale
in the sea, black. 'Leviathan, see!' said the driver.
'Jonah, maybe?' he laughed and the speedometer needle slid
higher, light flickering out in October dusk
like eyes, deep into night and ocean; we slid
like deep-water fish, or swam like the dead
over dark earth towards a mighty resurrection.
'Don't worry', said the driver, 'Nothing to be afraid of.
Halfway there now'.

Andrew Greig

ORKNEY/THIS LIFE

It is big sky and its changes.
It is sea all round and waters within.
It is the way sea and sky
work off each other constantly,
like people meeting in Alfred Street,
each face coming away with a hint
of the other's face pressed in it.
It is the way a week-long gale
ends and folk emerge to hear
a single bird squeal high up.

It is the way you lean to me
and the way I lean to you as if
we were each other's prevailing wind.
The way we connect along our shores,
the way we are tidal islands,
I'd settle for that. The way
you rise up by degrees on me, the way
I pick sand off myself later in the shower.
The way I am an inland loch in you
when a clatter of white whoops and rises.

It is the way Scotland looks to the South.
The way I go there and come back here again.
It is friends of my heart along the street,
the way we come and go through unlocked doors,
the way we enter each other's houses
to leave what we came with, or flick
the kettle's switch and wait.

This is where I want to live,
close to where the heart gives out,
perfected, a ruined arch against the sky
where birds fly through instead of prayers
while in Hoy Sound the ferry's engines
thrum *this life this life this life.*

Ian McDonough

A MACHINE WRITES HOME FROM THE ASYLUM

The clouds here frequently collide
   And clang like bells.
   This scares us
   As we have short memories.

On Wednesdays the Duchess comes
   To eat our language
   As it leaves our mouths.
   She is a fat lady.

Out in the yard
   Our keepers have us whistle
   At the shooting stars
   Which tear our sky
   In faithless, brief trajectories.

The dung of prowling Collietrons
   Is used to help our grass grow blue.
   They growl when someone learns
   To count past forty-five.

To-day the smell of sheep
   And craft-shop candles
   Was distilled for us to eat.
   The flatulence
   May last for years.

Cuckoo-clocks flap through the glen
   Drowning out the sound
   Of famous corpses shouting
   That they are not dead
   But only sleeping.

Send a postcard of Ben Badh.
   It lumbers through
   Our blackened dreams
   Trailing its roots
   Like an extracted tooth.

The clouds here frequently collide
   And clang like bells.
   We have short memories ...
   We have short memories ...

George Gunn

BLACK FISH

An orange fishing boat named Destiny
is tied up to the quayside
the village is its usual busy self
& the sun is shining
& the church sits high on the hill
to tell us what we do not understand
the bell rings 'Black fish, black fish'

now it's late at night & the pubs
are empty & closed, neon street lights
bathe the pier in a polythene filter
the Destiny has been joined
by the Pisces & the Shamal
a freezer lorry appears from nowhere
the men aboard the boats sing 'Black fish, black fish'

then the lorry is gone & the men are gone
& only the fishing boats remain
sitting like toys in a stone playpen
their bows as curved as rhino rib
their colours dulled by the envelope of night
& the village is asleep in its streets
dreaming 'Black fish, black fish'

A storm blows in wild deep brown
scuffing the firth into a frothing
tang of salt & the smell of the seabed
turned over like a bleaching sheet
after cattle, rattling the latch locked windows
& sighing as it hits the village
gable on 'Black fish, black fish'

from the Shetlandic

Robert Alan Jamieson

THE JAWBONES TALK

i) Kill

> ... the harpoon was darted: the stricken whale flew forwards; with igniting velocity the line ran through the groove – ran foul. Ahab stooped to clear it; but the flying turn caught him round the neck, and voicelessly as Turkish mutes bowstring their victim, he was shot out of the boat, ere the crew knew he was gone...   MELVILLE

Ocean shallows swell red –

ninety dead – ten swift minutes
for Sud'roy's matador grind-knifes
to have the bull-nosed heads detached,
revealing watermelon patterns,
and flooding juiceblood out,
the flow awash with dying.

men, women, children
wade the salty scarlet brew –
chattering joyous thanks, for this,
this manna fall from heaven.

here's meat for the storehouse –
oar straps, out of back-fin hides –
the stomachs, blown up, make good buoys,
boiled-down heads, the finest oil –
all clean-killed, in three quick days,

and all the waste dumped far to sea,
a fish meal, soaking slowly
back, into the ocean chain
to feed the future living.
no word of pain is spoken, just –

'The Lord is merciful and kind'.

ii) Krill

... most of the books about Iceland which I have read speak as if it were a nation of farmers. In point of fact, the majority live in towns, and pretty grim a town like Sigurfjordur is too... I see what was once a society and culture of independent peasant proprietors becoming, inevitably, proletarianised for the benefit of a few...　　　　　　　　　　　　　　　　　AUDEN

Nothing earth-bound's giant, on this dark isle.
Instead the winds are huge, the oceans massive –
There's a long way to go, to any where from here –
a small hope through that water, in this gale.

... the land is gathered; gathers from the shore –
bright-soaked timbers, dead Leviathan,
it gathers, great or little, what it can...

Comes a big ship, carrying stores, gewgaws, tools –
faces burn up, visions unimagined, never dreamt.
There's a long way to go, to get any thing up here –
but good things somewhere out there, after all.

... folk buy in supplies, they're wanting more –
an aluminium bucket, herring in a can –
wanting it in a tin, a tincan!

Comes a builder, carrying sheaf-bound paper plans.
Marks off land, and squares the round hill off.
It was a long way to work, at any job from here,
but now there's labour needed, back at home.

... workers steel the factory butchering floor
in Cannery Lane, and up the Fisherrow
no fishers row, oil motors roar ...

Nothing but whales can grow big here, no redwood
elephantine ostriches survive – only fat aquatic
capitalists, to suck the shining shrimpfolk in,
as with the ocean's soup, earth bonds her kin.

iii) Skill

INSCRIPTION, JAWBONE WALK, EDINBURGH: 'Presented by the Zetland & Fair Isle Knitters Association at the Edinburgh Exhibition, 1886'.

Between the jawbones of a whale,
the north isles knitters encamped
at Edinburgh's exhibition, erected
this symbol to represent their place,
set up shop and made good trade.

One hundred and eleven fresh years
have passed since they left this totem,
a tribal banner, gifted for posterity,
a standard planted 'in the Meadows' –
and still it hasn't grown a leaf.

But the passing world's coloured it green –
rucksacked students en route from elsewhere
to elsewhere bike, jog through the whale arch –
four curvilinear bones, knitted slightly skewiff
by a bolted iron cross, bridging them above.

And Reekie's traffic's turned these ivories
to trunks of trees, has camouflaged the
dislocation from their element of growth
with other foliage, other greens, till
bones? We see no bones.

Just the iron band around the base
to tell of folk and how they held
an exhibition of handiwork – made one
another gifts, and gave out ribands to
those with the quickest, cleanest skills.

These are the jaws of the gate to the nor'wastes.
Pass the mouth warily – here you might shiver,
mantle an ice dread. So call the dead knitters
to knit up a shroud, and skin these green bones –
they too were kin. Weep not for Ahab – he's gone.

iv) Skull

There was also a list: 'Jews to annihilate; Gothenburg'. X had been the man who was to be responsible ... He had made a thorough job. Even the names of the least well known, those who did not themselves know about their Jewishness, even those who thought themselves good reactionary Swedish bourgeois, as well as those of the best known, all the names were there. Jews to annihilate; Gothenburg... JAN MYRDAL

The bone has jaundiced –

All the soft tissue, rotted away –
the lips that were kissed,
that covered the flesh-tearing jaws
with such a sheepish grin, are gone.

Out of the mouth of the past,
we pluck the good white teeth,
bore them through and thread
the necklace of our history –
excluding those decayed
in long and grinding service.

The blackened stumps are left to rot
and those that crumble – rotten-hearted,
although appearing whole – are cast
to other people's mouths.

All the soft tissue of lies –
the lips that were kissed,
that covered the skin-piercing bite
with such a wolf-like grin – all gone.
Yet in the teeth, identity still
insists on silent talk – there's no

god or demon to choose herrefolk –
no one to lay the shame on, but
are of our ilk, their skull like mine.
None can explain destruction, with –

'The Devil is merciless unkind'.

'Thus I give up the spear!'

Jeffrey Shotts

SOLESCENCE

[Latin: *solus*, alone + *-escence*, becoming]

> Burned in this element
> To the bare bone, I am
> Trusted on the language.
> —W. S. GRAHAM

If it's not a word, it ought to be.
It drowns into itself –
Solescence, solescence, solescence–
Its liquid sequence becoming
Waves upon the quay beyond Aberdeen,
Near the cold Stones of Loanhead
And the cold North Sea, beside the autumn of the world:
That smell of burning,
Taste of dry leaves.

The standing stones, inland, hunch together
As if around a campfire and say,
*This is our language, at the edge of the world
And as old.*
The swelling sea swallows against the coast
As if chewing up the beach and says,
*This is our history, a devouring that is
Older still.*

In the delinquent reliquary
Between the brink of language
And the teeth of history,
A new word from an ancient vocabulary
Claims the sole essence
Of the stones
And the sea.

MORE BOREALI

IAN STEPHEN & MURDO MACDONALD

RONI HORN

GAVIN RENWICK

BARRY LOPEZ & ALEC FINLAY

WA-SHA-QUON-ASIN & FRIENDS

ALLICE LEGAT

EMILY CARR

IAN MCKEEVER

TONY MCMANUS

Vital Memorials

Remembering the Lewis Land Raids

Ian Stephen

IT SOMETIMES HAPPENS. A venture in the arts is innovative and challenging and yet touches the popular imagination. Maybe this unusual chemistry can only take place when the artist is in sympathy with the history and aspirations of a community, such as when, in Scotland, George Wyllie dramatises the demise of shipbuilding/engineering and asks where we're sailing now, or Tom McKendrick approaches similar subjects as industrial archaeology, with an arranged mythology and a reverence for the hard team-game which used to produce big scale works. Moving NW from the Clyde, recent works, designed by Will Maclean are now part of the terrain of Lewis – remembering the 'Land-Raids'.

Approaching from the South, clearing the Norse-style mountains of North Harris, Pairc cairn is the first. Close to the village of Balallan, it is settled on rising ground. The shape is an echo of a Pictish broch to be found 14 miles WNW at Dun Carloway. But the power implicit in this shape, realised in masonry work of outstanding quality, comes from more recent history.

The Pairc Deer-Raid of 1887 was a venison barbecue. People close to starvation entered the forbidden playground of the rich, to take food. It seems to have been almost festive, until arrests were made. Identified leaders were jailed but eventual judgments were lenient by the standards of the day. A symbolic aspect in the events – taking what was needed and was naturally provided – is marked by ritualistic elements in Maclean's design. Stones from the homes of the leaders are worked into the masonry. Stones taken from key-sites of the history are placed at the top edge, to guide the eyes of the viewer along the bearings of shores, hills and settlements. The shape is satisfying but also puzzling – you want to know where these protruding stones are pointing, and why.

Continuing N, through the commercial harbour-town of Stornoway, then Eastwards out along the isthmus which joins the Eye

[Photograph by Murdo Macdonald]

Aignis Cairn

[Photograph by Murdo Macdonald]

Peninsula to Melbost, Lewis, close to the site of St Columba's Church, two opposing walls now stand in permanent confrontation. At a proud height, each then curves back towards the dunes. Here at the site of Aignish Farm in 1888 land-raiders broke down the walls which kept them from enlarging their crofts up to the area necessary for subsistence. The treatment by the authorities was more brutal this time. It seems that establishment forces wanted to make their own drama, to issue a warning across the Highlands. Red-breasted marines who had been disembarked from HMS Seahorse and HMS Jackal rose up from the marram grass with fixed bayonets. The riot act was read. All this menace is carried into Maclean's design. Jagged stones on either wall are drawn against each other – whin branches against steel. The courts were also more savage this time. Here Maclean has made even greater demands on his collaborators in response to the life and death situation. They've responded. As with the Pairc cairn, ideas and vision had to meet structural tests and required exceptional masonry skills, to be made real. The initial designs were filtered by John Norgrove's engineering perspective. Then Jim Crawford, an artist who works in stone and mortar, carried out his commission under a stormblown tarpaulin.

Again the landscape has its own role. The site looks back SW towards the high deer-ground of Pairc. Down the open Minch, probably the route the warships would have taken, lie the Shiant

[Photograph by Murdo Macdonald]

Islands. In good visibility their own jagged drama provides another tense aspect. Then across Broad Bay, the machair at Gress carries the most Northerly of the three Land Raid memorials.

In the early 1920s, those who had returned from the insanity of French and Flemish battlefields (and who had escaped the fate of more than two hundred who perished on the Iolaire at the Approaches to Stornoway) claimed their promised allocation of land. Lord Leverhulme, the landowner, presented implacable faith in the mutual benefits of organised industry and land-use. But the crofters didn't want to be dairy-workers, merely holders of their own individual allotments. Maclean's composition of three stonework structures suggests something close to equilibrium in a locked confrontation of aims. Mounds of ground around these structures represent the trenches the crofters have escaped from. With that behind them, there could be no backing-down. To his credit, Leverhulme did give way, eventually, leaving his land to the people of Lewis to be run by an elected Trust.

As art-works, the cairns are special in the way that many boundaries are crossed, seemingly without effort. An artist with an international reputation (but with a Highland background) was happy to work with a committee formed to mark major events of local history in the most appropriate way. Local communities were involved throughout. Each opening ceremony was itself a performance event and the level of commitment was remarkable. Translate the percentage of people who marched to remember the Land Heroes, to a proportion of the Clydeside population and here is mass participation in the Arts.

The relationship between the artist who designed the memorials and the community is also remarkable and required a sense of vision from the original committee (Cuimhneachain Nan Gaisgeach, chaired by Angus Macleod of Marybank, Lewis). Maclean did not charge a fee for his involvement. The issue is one of his own major concerns. His relationship with Norgrove as engineer and particularly Crawford as mason is also worth noting, given the challenging nature of the demands placed on the one man who could lift the shape from the page.

Then there is the element of landscape art. Each setting and its outlook is visually accounted for in the designs. The history of each situation but also its geography is incorporated into the stonework. There was to have been a fourth memorial on the West side of Lewis at Great Bernera, but this time agreement with all parties was not reached. Yet given the emotive nature of the project, the completion of three memorials, each one innovative in design and realisation, is an achievement all involved should take pride in. Next, can we look back East to the Mainland and North to the big hills of Sutherland, think of that offensive statue to a certain Duke and explore imaginative ways of correcting its lies?

Ian Stephen is a writer and artist, based in Lewis. His most recent publication is *Broad Bay* (Morning Star, 1996).

Murdo Macdonald has recently taken up the chair of Scottish Art at the University of Dundee. He is currently writing a history of Scottish Art for Thames & Hudson.

untitled

Roni Horn

WHEN DICKINSON SHUT HER EYES

– *I go to Iceland* –

Recently I was reading the letters of Emily Dickinson. I began wondering about travel and Iceland too, wondering about the insistence of my returns here, about their necessity and the migratory regularity of them.
 I began to wonder about travel altogether, about the how and the what of it. Travel isn't so simple as a car or a train, or as nameable as a place. I thought about Emily Dickinson's travels. From the first letters she wrote she told her correspondents she didn't go out, she didn't want to go out, and that she would not come to visit them. Dickinson stayed home, insistently. Locking herself into her upstairs room, she invented another form of travel and went places.
 Dickinson's invention was multiplication, herself and empirical reach: everything that could be felt, heard, seen or smelt, everything perceptible, everything discernible from 280 Main Street, Amherst, Massachusetts. Perceptible includes the library; somehow Dickinson used the library as an empirical source, somehow she learned to consume its contents sensorially. Her library was not a source of acquired knowledge, not a tool of the intellect. Her library was simply another perceptible thing becoming another entrance, confirmation of all she sensed in the world. Even her poems about God and being dead are eyewitness.
 Sequestered from the world, knowing that going out into the world hampered her ability to invent it, Dickinson stayed home except for two summer trips to Cambridge as a child and one to Washington later in life. Dickinson stayed home when Ralph Waldo Emerson visited her brother next door. Her business was circumference.
 In her verse Dickinson spoke of Vesuvius at home. In her letters she said she travelled when she closed her eyes, and that she went to sleep as though it were a country. In her room alone, she said, was freedom. Here she wrote one thousand seven hundred and seventy

five poems. Dickinson shut her eyes and went places this world never was.

For the time being, Dickinson's here with me—in Iceland. For someone who stayed home she fits naturally into this distant and necessary place. Her writing is an equivalent of this unique island; Dickinson invented a syntax out of herself, and Iceland did too—volcanos do. Dickinson stayed home to get at the world. But home is an island like this one. And I come to this island to get at the very center of the world.

SOMETHING SHIMMERING

Outside the sky is vivid and bright: it's blue, and the brilliant white clouds snap with orange. Their unusual shapes provoke no metaphor. Here cloud-watching is not daydreaming. Cloud-watching takes you to the heart of things, to a deeper, more elusive reality.

Each cloud settles into a nameless composure taking the shape of itself. That self is that particular cloud floating in that particular place and in that moment. It's that and nothing more. Sometimes—like tonight—the sky is central and the landscape is merely a surface upon which the shadows of clouds gather. The clouds bundle together, cramming into one another, the occasional one floating lone and free. When you can't sleep around here it's because something's happening in the sky.

Once, years ago, I was lying restless and awake in my tent. Past midnight I stepped outside. A full moon, I thought, would confirm an astronomical influence. But here in Iceland it wouldn't simply be the moon, it would be something else, something less discrete. The planets attracted to my view were converging on my gaze: they gathered outside my tent.

How else could you explain what I saw? To look at it, I discovered no object but a tremulous something. This something shimmered; that's all I could make out. There it was: a fitful shining, luminous and brief.

The planets gathered in my view, called together by my gaze. My gaze, called out by the view, called out by the planets as they composed the view, differently, in gathering. This is cause and effect.

ROADS LACK DEDICATION

A path. Drive it, walk it, take it to be somewhere. When you're on a path, you're in the place you're in. There's no distinction between the path and the place itself. The path's a minor clearing, a simple parting. But it's a complex thing because it takes the shape of each place, intimately. A path in sand is more sand, sometimes rustled up; a path in earth is compressed earth; a path in stone is a slight leveling of the stones. Sometimes a path is nothing but an indication on a map, the reality having been blown or washed away. The path goes through places usually the most beautiful way so that when you're on this way you can see beautiful things. Sometimes the path wanders about and you wander about with it.

Direction lends order and tempo. When you move along, everything on the move is part of the path, as well as the way to another place. The path can only be where it is and nowhere else: the path is dedicated.

A road is not a place. A road is a platonic surface homogenizing with velocity and predictability. A road takes you from here to there and what's in between is merely in between. On the road, the thing is to get there. The road clears everything out of the way, gets you there no nonsense. You're nowhere but on the road when you're on a road; you're not where it is. It's the nature of the road, it takes you out of the place you're in and makes the road the where you are. It's the difference in form between the road and the place it's in; it's the potential for speed and the linear inclination. Sometimes the road is a vector, in Texas or on Skeidarársandur for example.

Up through 1985 or so, Iceland had nothing but a country-wide network of paths, which if you were committed enough, mostly got you where you were going in time and without ever removing you from the landscape. But now Iceland's largely been converted to roads and you're rarely where you are when you're on them. Roads lack dedication. That means making the island a place like any other.

HER, THE WATER, AND ME

I've been watching her in the water—in the pond, in the cool pond water. These slow Sunday summer afternoons she takes her time in the water, and I watch. I watch as she soundlessly bobs, her head floating above, her body weightless and cool below. Occasional ovals of her whiteness surface briefly and then submerge, making sounds of muffled fluid. The water with her, approaches me, as I sit upon the shore. The heat tickles at my face as I watch. The sweat slips on my skin as I watch. My eyes, unblinking, hold the view in the bright invasive light.

The sky diced in myriad tiny fragments lies about her. Carried on the rippling water, the diced brightness radiates from her bobbing head, quietly covering the pond. The trees above steady my gaze. Their reflection breaks upon her and mingles with pieces of sky.

As she floats, the water commingles us in the world. Its unseen cool darkness places us, holds us, an entrance easily recognized from all over.

Water carries her image from her. Water carries her portrait to the edges of the pond. Water on her skin—drips, drops—moving over the pores and the freckles, each drop slowed, elongated by the drag of her flesh and all the other drops of her sweat.

Reflected light and wetness on flesh make her freckles pop. Wetness sits lightly among her almost invisible hairs—an aura of dampness and heat—imparting her skin depth and vividness. Drips-drops, each one magnifying and crystallizing moments of her flesh. Incidents of her skin appear enlarged. Little monuments sliding and evaporating a way to invisibility. Water in her hair bundles each strand with weight tight against her head. Her dark hair and the lapping water about her teem with the unchanging brightness of the day.

Water carries her image from her. It infuses the trees, the clouds, and the shore. She surrounds me. But water is only the half of it—the other half is the moon, the face, or the evergreen. The other half is the friction and the gravity. The other half is us.

Roni Horn is a visual artist. She lives in New York and often sojourns in Iceland. 'When Dickinson Shut Her Eyes' was first published in *Pooling Waters* (Walter Konig: Köln, 1994). It is the fourth volume in the work *To Place*.

More Borealis[1]
a Canadian travelogue
Gavin T. Renwick

TRANSATLANTIC

Above the ice, crossing the threshold between sea and land, between winter and spring, brings to mind that mythical place beyond the northern wind where all northern peoples were thought to belong originally – *Hyperborea*, the necessary antithesis to Grecian 'myopic' light. And swept around this North Atlantic arc is a migratory dynamic from Celtic monk and Nordic settler to cleared Highlander. A considerable part of Scottish history was lived in that landscape below (a wasteland to southern eyes?), where we met half-way a North American kindred with common archaic Asiatic ancestry.

THE TERRESTRIAL TRACK IS FOUND

75 degrees in April heightens the sense of transition between worlds. The sheer scale of activity creates a sense of intellectual and physical liberation, sitting around New York's cosmopolitan hearth. It seems the concept of *America* began – and ended? – here. Beyond lies the nation state – legislated space and expropriated land.

On the way to somewhere else, through the fledging colonial outpost that was New England, a reflective rail journey follows the Connecticut river to an American end. Transcending other colonial frontiers, Montreal's toponymy reveals an Auld Alliance in another 'new' world:

RUE MCTAVISH

Proximity to the St. Lawrence River enriched both this city and its Franco/Celtic founders. Another later type of Scottish emigre lived in bourgeois gentility here, Montreal being the fulcrum of the North West Company's business, and the route through which the produce of the Nor'West trappers passed on its way down the St. Lawrence to grace the heads of Europe – in turn, the delta was the conduit into Canada for the ideas of nineteenth century Empire.

The North West Company was fated to be absorbed by their older, state-sponsored (crown and parliament) rivals, and its Highland-emigre partners became employees in an English-based corporation of 'gentleman adventurists'.

As if evidence of Victorian appeasement, Ottawa sits on the bank of the river that gave it name, diplomatically sited between the opposing cultural capitals of Upper and Lower Canada. Teasingly, the House of Commons, placed on the British side of the divide, consciously turns its back on Quebec, as the urban fabric of Ottawa's sister city of Hull on the opposite bank now does in return. But directly across from the parliamentary Gothic, architect Douglas Cardinal has designed a museum as an analogous ring, conceived to unite the 'new world' dichotomy. Exhibited within this celebration of Canada is the architecture of many cultures, a complexity united by the organic modernism of the building as a whole – an appropriate if idealistic national allegory. The containment of the vernacular by this modern state-built structure can be seen as a metaphor for Trudeau's homogenous, and ahistorical, liberal vision of federal Canada.

AN IMMENSE EXTENT OF COUNTRY

Passing endless Toronto suburban housing echoing generic Euro-American models that maintain a passive uniformity despite contextual and climatic change, my route follows the trans-Canada through tired Ontario towns held in hibernation by Lake Superior's April ice. It engenders a wintry negativity, exaggerated by signs:

CLOSED FOR THE SEASON

Compressed evolution has meant that 'boom-town' and 'downtown' co-exist in recent living memory. Inspired by the prime ministerial vision of a pan-continental Canada, a native of Sutherland, Sir John A. Macdonald, tied the coasts together with a railway. The Canadian Pacific would on completion transport through these towns the commodities of the fledgling west – for so long, the wealth of Canada had been concentrated in the St. Lawrence valley. This was complemented by the exportation 'out there', to the advancing agrarian frontier, of the products of the Valley's manufacturing industry – so to, abetted by the colonisation companies and the homestead policies integral to the railway's development, went notions of society and home.

Descending the Canadian Shield towards the Manitoban border, the metropolitan dominance of Montreal and Toronto and their Anglo/Franco animosity fades. Leaving that part of history behind, my route enters what was formerly Rupert's Land, an ecosystem imperceptible in scale in London at the time of its notional delineation. The 1670 charter, championed by the then monarch's son, defined all the land draining into Hudson's Bay as the colonial 'corporate' holding of 'The Governor & Company of Adventurists trading into Hudson's Bay', fully intending that the company be 'absolute lords and proprietors'[2] over lands granted. The problem lay in how the imperial mind interpreted this space:

> Imagine an immense extent of country, many hundred miles broad and many hundred miles long, covered with dense forests, extended lakes, broad rivers, wide prairies, swamps, and mighty mountains; and all in a state of primeval simplicity – undefaced by the axe of civilised man, and untenanted by aught save a few roving hordes of Red Indians and myriads of wild animals. Imagine amid this wilderness a number of small squares enclosing half-a-dozen wooden houses and about a dozen men, and between each of these establishments a space of forests varying from fifty to three hundred miles in length; and you will have a pretty correct idea of the Hudson Bay Company's territories.[3]

PARALLEL LINES

Entering 20th century Canada, following the direction if not the experience of the early emigres, it seems this trail and the parallel railroad have helped to formulate a contemporary linear nationality. Hudson's Bay and Nor'Wester posts, mounted police forts and railway stations initiated ethnically diverse pioneer townships, which rapidly consolidated into a highly urbanised population, mostly inhabiting a seventy miles wide corridor along the border. This area of clustered development contains 75% of the population, all proudly Canadian, yet all with one eye looking south, both enticed and appalled by what they see – a situation which many Scots identify with. A national geometry emerges on the Greyhound journey, between coffee and doughnut shops in apparently horizontal towns – the effect of flatness exaggerated by ploughed snow floodlit by parallel streetlights. With some of these settlements nearer to the American border than to each other, you begin to appreciate how tenuous the federation might be. Yet the common experience of western migration along this trail into legislated landscape is seemingly strong enough to resist the southern treasures.

HIGHLAND PIONEERS

Passing through this 'empty' landscape, occasional roadsigns suggest places meaningful to different tribes: War Eagle Lake, McCallum Point. After a tortuous northwestern journey from

forced clearance in faraway Caithness, the Kildonan Highlands arrived at this Manitoban Red River point as Selkirk's pioneers, supported by the Hudson's Bay Company. To these first settlers, the circumstances must have seemed familiar, similar to the feudalism back home. They found themselves once again virtual vassals – only now their loyalty was nominally to the Company, not the clan.
The 'Bay always presumed that being lords and proprietors of the soil gave sovereignty over all trade. This autocratic attitude obviously created conflict if an area was already being trapped, as with here. The Red River pioneers were caught in a corporate battle for trading supremacy, on the 'Bay's side, against a Metis resistance encouraged, ironically, by a MacGillivray, the emigrant son of a Jacobite family with allegiance to the locally established North West Company.

Alexander Ross was a Nairn-born Nor'Wester who worked for the Hudson's Bay at the Red River settlement after the amalgamation of the two companies. A previous fortuitous meeting with the formidable post incorporation Governor, Ullapool-born George Simpson, resulted in a land grant slightly upriver from the colony, in what is now downtown Winnipeg.

Surveying the contemporary vista from the same site, it still has the feel of a frontier town. The Victorian morphology of municipal and railway architecture, the verticals of downtown office blocks create a centrality, in defiance of the liberating prairie landscape. Despite deep snow and intense cold, the city operates. Unlike Ontario, the roads are snow-covered but dry – the authorities here use sand as opposed to salt. The absence of saline slush illustrates a different perception of winter, and I feel as if I have entered another Canada.

PRAIRIE FEVER

Mechanical rigour is all that determines perceptible movement westwards, sequential tail-lights hint at an indefinite distance ahead. The radio waveband's empty other than vague voices struggling through the space between. Entering Saskatchewan, a 'donated' warning of monotony comes to mind. But the prairie is such a contrast to Scotland that it raises a fascinating, formal problem – what must it be like to occupy the prairie? The image of oneself as an anomaly in this uncluttered plane/plain is the solitary dynamic vertical in a horizontal betrayed only by horizon:

> I have thought myself home,
> run through immeasurable space
> ahead of time.[4]

FROM CALGARY TO CALGARY

Moose Jaw, Swift Current and Medicine Hat are passed. Having followed two setting prairie suns to the last mid-west province of Alberta, the plain ends abruptly. The youthful arrogance of the Rockies is the first north-south axis since the Red River as, backed by the mountainscape, Calgary becomes evident at the threshold between topographies, a locus for reference lines of mountain, river and rail.

On the wall of Calgary's City Hall is a mural interpreting the bay on the Isle of Mull to which this newest of North American cities owes its name. The language of the people who once inhabited that cleared bay is enshrined in the University motto, its crest echoing Clan MacLeod's. Crucially, the University houses a faculty for general studies, where shared epistemologies create for scholars a place in the Scottish tradition of democratic intellect.[5]

Twenty four hours trajectory steadily northwards is halted by the mighty Mackenzie River – the name of a Stornoway native gracing this most evocative boundary. At this intersection, the north-bound river teases by flowing westwards, as if trying to lead the Pacific bound explorer astray, 'beyond the sphere of European influence'.[6]

'I sat up last night to observe at what time the sun would set, but found that he did not set at all.'[7]

A tentative voice suggests that the ferry might operate. Our arrival had coincided with the break up of the Great Slave Lake ice sheet. But with trepidation, the time-served vessel successfully negotiates the flow of fractured ice.

Parallel histories and perceptions are reflected in the place-names of the northern region. The Englishman Franklin, representing state and admiralty, on his way to a 'heroic' death, was followed by the Scotsman Rae, representing the Hudson Bay Company, on this way to find him. The former symbolises the astounding pretension of Empire; the latter displayed deference to the Company, while harbouring an intuitive respect for aboriginal society originating in his Orcadian upbringing – a relationship of personalities not dissimilar to that of Scott and Bruce at the other polar extreme. Conflicting interpretations of the same colonial frontier notwithstanding, both were to leave their surnames on the mapped land.

THE NORTH CAN TELL US EVERYTHING THAT CANADA HAS BEEN OR WILL BE[8]

Despite its demographics, the iconography of Canada is largely northern. Yet most 'euro-Canadians' are less than at ease with their native compatriots. My point of arrival in the western sub-Arctic –

the territorial capital of Yellowknife – reflects in microsm Canada's multifaceted history, for better and for worse. Its political and commercial centrality to a vast region and its proximity to first nation communities necessitates a sharing of public space which in southern Canadian towns might be unconsciously segregated.

Like those Gaelic names gradually thinned from British Ordinance Survey map reprints, the long established Dogrib Dene trail linking river and lake ecosystem between the Great Bear and the Great Slave has no cartographical presence, but the Dene names of the landscape (both sacred and profane) still exist in a precarious oral tradition. The transmission of this knowledge generates a sense of intimacy with place and if names are indeed 'an extraordinary potent starting point for a revolution'[9] as one newspaper commentator recently phrased it, then it is through the discarding of 'Fort Franklin' and 'Rae Lake' that the inhabitants of 'Deline' (*moving or flowing water*) and Gameti (*rabbit-net lake*) begin to reclaim their homeland. Thus, as Celtic peoples rediscover multi-centricity in a pan-continental nation stretching from Briezh to Cape Breton and unite to challenge established conceptions of a monoculture, so Canada – be the reason economic, political or cultural – can now challenge the acceptable convention of a single south-eastern centre. The current prominence of Calgary, the symbolic importance of Yellowknife and the emerging Baffin Island capital of the Inuit territory of Nunavut – Iqaluit – are all examples of this growing multicentricity.

A JOURNEY IS A LINKING IN SPACE

Two voyages out from Yellowknife link end of the transitional zone between ecosystems of *taiga* and *tundra*. The first ice free weekend allows, by Hudson Bay canoe and mosquito-filled portage, a sojourn at a hidden loch. Our heavy footsteps followed indentations pressed in the snow by a winter of skidoo travel, till they disappeared at the newly liberated lake. Here the ravished geology of the northern shield was so beatified by archaic sphagnum and glorious lichen, I felt again that sense of elemental liberty last experienced on the prairie. A ritual birthday toast of *uisge beaha* later, and something presbyterian about the purity of the paper birch trees sheltering the rear of the campsite seemed to generate a warm symbiosis, subtly absorbing and reflecting the surrounding colour, a oneness.

And on a second journey, by Twin Otter and boat to a Dogrib Dene summer camp, I am requested – honoured – to thank my true host, the land and the lake. Later that evening, on a boat trip with a Dogrib Elder to check his nets, I recalled a trip I made from the shores of Mull with a comparable 'elder gentle man' over there. I felt a sense of other cycles impinging from beyond the visible timeframe, and a commonality between these custodians of the earth and their respective homelands. As the shallow sun and long shadow of winter seemed to show the contours of the footprints around ancient shielings in the Gaidhealtachd, so here in Denedeh the dry air and intense cold appeared to preverse centre-poles that once gave structure to ancient tipis. The problems associated with continuance of the local way of life in the 21st century seem common to both Dogrib and Gael. Straddling different world-views must be a daunting task for the young. But here in the North West, a 'ceilidh' with drums and dancing around the hearth[10] confirms the succour to be had there.

Over both journeys spiralled the Arctic Terns, newly arrived from Antarctica, living a seasonal constant thanks to bi-polar migration. These chance encounters concluded a balancing of my perceptions of 'world' and 'home' between that moment standing on the sand of Calgary Bay, Mull, facing Canada in the company of the Gaelic elder, and this, on Dogrib land – like the tern linking in space, between the two poles of 'cosmos' and 'hearth',[11] a common but ancient philosophy.

NOTES:

1. A creative expansion of the Latin phrase 'more boreali', adopted as the name of an informal, interdisciplinary organisation dedicated to an expanded understanding of Scotland as part of a boreal community in the north.
2. Ross, Alexander, *The Red River Settlement* (1856, London; Facsimile edition 1957, Minneapolis), p. 2.
3. Ballantyne, R.M., *Hudson Bay* (London, 1912), p. 52.
4. From 'Thinking Home', written in 1874 by Icelandic poet and Albertan settler Stephan G. Stephansson, included in *Selected Prose and Poetry* (Red Deer, 1988).
5. Davie, George, *The Crisis of the Democratic Intellect* (Edinburgh, 1986).
6. Hunter, James, *A Dance Called America* (Edinburgh, 1994), p. 154.
7. Daniells, R., *Alexander Mackenzie and the North West*, pp. 46-49.
8. Hume, Stephen, 'A North that can be felt but never known', in *Knowing the North* (Edmonton, 1955), p. 100.
9. Kettle, Martin, in 'The Guardian' (10.5.97), p. 21.
10. 'Hearth' is the literal translation of 'kǫ', Dogrib for 'house'. It can also mean 'fire'.
11. Tuan, Yi-Fu, *Cosmos and Hearth* (Minnesota, 1996), p. 14.

Gavin Renwick is a designer & researcher affiliated to Napier University, and a partner in the multi-disciplinary collaboration *Whaur Extremes Meet*. He is currently working between Scotland and Canada.

'Wa-Sha-Quon-Asin' & Anahareo

Notes from Grey Owl's Trail

1. ... THIRTY MILES BEYOND Waskesui and accessible only by water and portage lies lovely Lake Ajawaan – insignificant in terms of size but in the 1930s the most widely publicized and best-known body of water in Saskatchewan. For to the lake in 1930 came that remarkable half-breed Grey Owl to take up his duties as conservation officer for Prince Albert National Park. Grey Owl, who said he was the son of a Scottish father and an Apache mother, brought with him to Lake Ajawaan his beautiful half-breed wife Anahareo; a passionate belief in the need to protect the North American beaver from extinction; and a growing reputation as a writer of fascinating true stories about wilderness life and wilderness animals ...

2. ... This is primarily an animal story; it is also the story of two people, and their struggle to emerge from the chaos into which the failure of the fur trade, and the breaking down of the old proprietorial hunting grounds, plunged the Indian people, and not a few whites, during the last two decades. Their means of livelihood destroyed by fire and the invasion by hordes of transient trappers and cheap fur buyers, these two, a man and a woman, newly married and with no prospects, broke loose from their surroundings taking with them all that was left to them of the once vast heritage of their people, – their equipment and two small animals as pets.

Outcasts in their own country, wandering what amounted to a foreign land, they tried desperately to fit somewhere into this new picture. Their devotion to these creatures that represented to them the very soul of their lost environment, eventually proved to be their salvation ...

3. ... Grey Owl was born in 1888 of Scotch and Indian parentage. He went to England for a while, and returned to this side quickly thereafter, taking part in the Cobalt silver rush of 1905. He was then, as he has been ever since (but for the space of his War Service), a canoeman and packer. He never forgets his great debt to the Ojibway Indians. He was still a youth when he was adopted into their tribe. It was they who named him Grey Owl because of his habit of nocturnal travelling. He learned their language. From them

he derived his forest lore. He lived their nomadic life ...

4. ... Many years ago I cast my lot in with a nation known under the various appellations of Chippeways, Algonquins, Londucks, and Ojibways. A blood-brother proved and sworn, by moose-head feast, wordless chant, and ancient ritual was I named before a gaily decorated and attentive concourse, when Ne-Gank-abo, 'Man-that-stands-ahead,' whom none living remember as a young man, danced the conjuror's dance beneath the spruce trees, before an open fire; danced alone before a sacred bear-skull set beneath a painted rawhide shield, whose bizarre device might have graced the tomb of some long-dead Pharoah. And as the chanting rose and fell in endless reiteration, the flitting shadows of his weird contortions danced a witches dance between the serried tree-trunks. The smoke hung in a white pall short of the spreading limbs of the towering trees, and with a hundred pairs of beady eyes upon me, I stepped out beneath it when called upon ... the sensation of stepping into a motionless ring was that of suddenly entering a temple, devoted to the worship of some pagan deity, where the walls were lined with images cast in bronze; and there proudly received the name they had devised, which the old man now bestowed upon me ... Wa-Sha-Quon-Asin: Grey Owl.

4. ... Early becoming one of them, he feels that they are his people and that all he is and has he owes to them. They taught him to love Northern Ontario and to think of it as his home land. Their land was his land and their folk his folk. The Indian influence, or rather the Ojibway Indian influence, is naturally very marked in all his reminiscences and portrayals of wilderness life.

5. ... In order to properly grasp the spirit in which this book is written, it is necessary to remember that though it is not altogether an Indian story, it has an Indian background. The considering attitude towards all nature which appears throughout the work, is best explaining by a quotation from John G. Gifford's 'Story of the Seminole War.'

'The meaning of sovereignty is not very clear to primitive peoples, especially to the Indian. He rarely dominated the things around him; he was a part of nature and not its boss.' Hewitt says of the Indian:

> 'In his own country ... he is a harmonious element in a landscape that is incomparable in its nobility of colour and mass and feeling of the Unchangeable. He never dominates it as does the European his environment, but belongs there as do the mesas, skies, sunshine, spaces and the other living crea-

tures. He takes his part in it with the clouds, winds, rocks, plants, birds and beasts, with a drum beat and chant and symbolic gesture, keeping time with the seasons, moving in orderly procession with nature, holding to the unity of life in all things, seeking no superior place for himself but merely a state of harmony with all created things ... the most rhythmic life ... that is lived among the races of men.'

This viewpoint is not peculiar to people of native blood but is often found in those of other races who have resided for many years in the wilderness.

6. ... A canoe is to Grey Owl what a horse is to a cowpuncher or a good vessel to a sailor. Prior to his becoming so deeply interested in what has turned out to be his life work, namely the preservation of the Little People, his days were spent in guiding, exploration and transportation of supplies up and down and across and about the north country. He trapped every winter, and for a few summers served as a forest ranger for the Ontario Government. He was singularly successful, and his ability to penetrate easily though unexplored territory gained for him a roving commission. The war stopped his activities for three years. He returned from it pronounced unfit from wounds in 1917 ...

7. ... He made his headquarters in Bisco before and after the First World War. When he came back from overseas he was drawing full pensions, which gives you an idea of the state of his health at that time. He had been in a gas attack and had shrapnel in his foot. As a badly wounded soldier with little hope of recovering, he faced a futile and unhappy existence. He became dispirited and gloomy and turned to music for solace. He used to buy records by the dozen and take them to a friend's home to sit in the darkened parlour with his special brand of scotch and his memories, listening to music by the hour ...

8. ... As soon as he was able he resumed his former manner of living. His speed and endurance and extraordinary intimate knowledge gained for him the post of assistant Chief Ranger over a large area in the Mississauga Forest reserve. After a few years during which he came to know every nook and cranny of this region to his own satisfaction a ranger's life became monotonous and far-off horizons beckoned. He closed his old trapping camp on the Spanish river, threw together a light outfit and set out on new wanderings, hiring out, canoe and man, wherever guiding, packing, and the like provided means of renewing supplies. He tramped over a new hunting ground every winter.

9. ... It was late in the summer in 1925 and I was at Wabikon, a resort on Lake Temagami. I was reading in the shade of the pines, when I was interrupted by a gritting sound from the beach. Looking up to find the source of the disturbance, I saw a man dressed in brown deerskins stepping with the speed and grace of a panther from a canoe. And there he stood, tall, straight and handsome, gazing wistfully across the lake in direction from which he had come ...

His shirt and trousers were dark brown, brightened by a Hudson's Bay bely and a much worn buckskin vest, which matched the moccasins on his feet. But what really set my imagination afire was his long hair and wide-brimmed hat ... this man looked like the ever so thrilling hero of my youth, Jesse James, that mad, dashing, and romantic Robin Hood of America ...

10. ... We were in many ways, exact opposites ... My Gertrude, who will be referred to from now on by her tribal name of Anahareo, was not highly educated, save in that broader sense which is much to be desired, and is not always the result of schooling. She had a passion for advancement, uplift, and a proper use of words and took lively interest in world events. She was direct descendant of hereditary Iroquois chiefs, and her father was one of the original Mohawk river-men who had helped to make history along the Ottawa in the days of the great square-timber rafts; she came of proud race. She was strictly modern, as modern went at that time, a good dancer and conversationalist, and a particular dresser herself, she naturally wanted me to always look my best ...

... I speedily discovered that I was married to no butterfly, in spite of her modernistic ideas, and found that my companion could swing an axe as well as she could lip-stick, and was able to put up a tent in good shape, make quick fire, and could rig a tump-line and get a load across in good time, even if she did have to sit down and powder her nose at the other end of the portage. She habitually wore beeches, a custom not at that time so universal amongst women as it now, and one that I did not in those days look on with any great approval ...

... But the exigencies of constant travel and the demands made by our manner of living, left little time for temperamental readjustments and the hurry and of early winter trapping occupied the full of our days, until one winter evening just before Christmas I returned from a short trip to find my proud and gallant Anahareo in a dishevelled and disconsolate heap upon the bunk ...

... my point of view was slowly changing. Forced at last to stop and look around and take stock, obliged now to think of someone else besides myself, I stepped out of the case hardened shell and rubbed my eyes to get a clearer vision and saw many things that had hitherto escaped me ... I began to have a faint distaste for my bloody occupation. This was resolutely quenched, though the eventual outcome was inescapable ...

11. ... Then came the moment of the conversion, from destroyer to preserver, when the whimpering of two orphan beaver kittens whose mother he had just shot impelled Grey Owl to swear a solemn oath never to kill another beaver. He and Anahareo raised the orphans by bottle and thereafter became increasingly devoted to the cause ...

... in sparing the beaver Grey Owl had cut himself off from what had for long been his chief means of support. But he persevered in the beaver cause: a few articles he wrote on the subject of conservation for an English magazine proved popular enough to warrant the publication of a full length book, *Men of the Last Frontier*; the book caught the eye of the National Parks Service and led to Grey Owl's appointment to the conservation post in Prince Albert National Park ...

11. ... The winter was not without event. The book was accepted, and it had been published practically verbatim, except the title, which had been altered ... I received a number of press notices. Most reviewers were kind and let me down easy, even praised me ... One or two critics ... mildly scandalized apparently that an uncultured bushwacker of acknowledged native blood should step so out of character and become articulate, were more severe. They seemed to take it as a personal affront that there were in existence beings who, without benefit of education, had common knowledge of many things not taught in halls of learning, casting, by implication, some doubt on my knowledge of a subject with which they could have had but little acquaintance; quite as though they were, by some divine right, omniscient. They probably did not realise that the rather standardizing influence of an intensive education militates somewhat against the development of an ability to grasp the more subtle and elusive nuances of a culture peculiar to the wilderness ...

12. ... Thereafter life closed swiftly in. Grey Owl wrote more books, all of them best-sellers; he fought a successful battle for beaver conservation; he became the most widely-known Canadian author and lecturer of his day. Twice he toured England, lecturing to large and fascinated audiences. He was received by King George VI and the Royal Family, in whose presence he conducted himself with extraordinary poise and dignity. Everywhere he went, he carried

himself, not like a half-breed, but like a great chief. He looked the part – tall, lean, bronzed, hawk-nosed. And acted it – proud to the point of arrogance, quick to resent the faintest hint of condescension on the part of the white man ...

13. ... The idea of domination and submission, though now passing out of date in nearly every walk of life, is hard to disassociate, in the minds of some, from the contact between civilized man and beings in a state of nature. This was forcibly illustrated in a late radio broadcast during which, in a play dealing with frontier conditions, an actor who portrayed the part of Indian guide was heard to address the head (not the leader, as is generally supposed, the guide being of necessity in that capacity) of the party, in awed voice, as 'Master.' But the more tolerant and unaspiring, though perhaps less ambitious viewpoint of the Indian must be taken into consideration, if the reader is to fully appreciate the rather unusual tenor of the narrative ...

14. ... The wilderness man adjusts ill to urban society, and the pressures mounted. Alcohol eased them, relaxed Grey Owl to the point where getting him pulled together for his platform appearance became a major concern of his attendants. Still, once on the platform he told his story well. No irrelevancies, no aberrations. Except, perhaps, that evening in Hastings on the Channel coast when he invited anyone in the audience named Belaney to come and speak with him ...

... The next year Grey Owl died in a Prince Albert hospital, his ravaged body unable to withstand the shock of pneumonia. Animal lovers around the world lamented the news of his death. And a newspaper man in Hastings remembered an odd word spoken – the word Belaney.

15. ... The controversy about Archie's origin went on relentlessly in the weeks after his death. If by this time I had any ideas left about Archie being from Mexico, they were thoroughly shaken when I read the newspaper account of his mother's letter to Miss Ada Belany (who was her sister-in-law, not her sister as the newspaper said).

GREY OWL'S 'MOTHER' WRITES TO SISTER ON FAMOUS 'SON'S' DEATH
LONDON, April 21 [1938] – The British Press today continues to interest itself in Grey Owl, the naturalist who died last week in Saskatchewan where he made his home. The question is whether he was an Indian half-breed, as he maintained, or Archibald Belany, an Englishman with no Indian blood.

Latest development is a letter published in the News Chronicle as one sent by Belaney's mother from Devon to her sister at Hastings.

'I write with regret of the passing of my beloved son' the letter said. 'It is the passing of a great man with a soul so unique that never has been another like him.'

When finally I was convinced that Archie was English, I had the awful feeling for all those years I had been married to a ghost, that the man who now lay buried at Ajawaan was someone I had never known, and that Archie had never existed ...

16. ... The truth ... affirmed that Grey Owl was without a trace of Indian blood, was in fact a Hastings-reared Englishman, Archibald Stanfield Belaney. He had come to Canada at the age of fifteen; he had found his way into the Ontario north woods and had become so enamoured of the Indian way of life that as a preliminary step to becoming wholly absorbed in it he had severed all connection with his own people.

Inevitably those persons most intimately associated with Grey Owl in the time of his reknown – including his English publisher who had camped with him in of all places Epping Forest on the outskirts of London and listened to a long harangue on the plight of the half-breed – at first refused to accept the implications of the revelation. Others did so, but fiercely resented having been hoodwinked. Grey Owl, whom they had loved for his humanitarianism, his noble chieftain's mien, his romantic aura, his fine tales, was a fake, an imposter. An Englishman named Belaney ...

17. ... The revelation of Grey Owl's true background and original name brought widespread surprise and interest. But what was far more important was the message brought to Canadians and others of wild things having rights and deserving to be treated with compassion. He had been a trapper but thanks to the influence of Anahareo ... he abandoned trapping and all the needless killing. 'I realized,' he wrote, 'what a crime we trappers were committing against nature that had been so bountiful. I dedicated my life at that moment to conservation of game.' The wild creatures became his friends, and he found an unusual talent for communicating with them. They seemed to trust him and he did not fail them.

18. ... I have spoken of him as being almost missionary in his quality. He is. The cause of preservation and conservation of the Wilderness and its folk is his lifework, and he feels himself as surely called to it as a man of the cloth is called. He might paraphrase John Wesley and say, 'The Wilderness is my parish.' In Grey Owl's own

words, 'Give me a good canoe, a pair of Jibway snowshoes, my beaver, my family and ten thousand square miles of wilderness and I am happy.' He does not add, but I may for him, that he has in ample measure another requisite for happiness: he is a happy man because he has learned to help others to happiness, and amongst those others not least, his friends the Little People.

19. ... In the rather ill-conceived rush we have been in to exploit our natural resources, we have taken little trouble to examine the capabilities and possibilities of the wild creatures involved in it, save in so far as the findings were of commercial value. Therefore much that is interesting has been overlooked. The kinship between the human race and the rest of our natural fauna becomes very apparent to those of us who sojourn among the latter for any length of time; alarmingly so to those whose attitude has hitherto been governed by the well-worn and much abused phrase that "Man shall have dominion over all." However I do not draw comparisons between man and beast, save in a few instances which are too remarkable to be overlooked. Nor do I ascribe human attributes to animals. If any of their qualities are found to approximate some of our own, it is because they have, unknown to us, always possessed them, and the fault lies in our not having discovered sooner that these characteristics were not after all exclusively human ...

20. ... The fact that Grey Owl was an Englishman named Belaney ... in no way affects the value of his achievement. His work as a conservationist, writer and publicist on behalf on the Indian and the half-breed is not degraded by the revelation of his identity. Only for those who can never separate the work from the man, who damn Burns' poetry because Burns drank and rate Shelley below Tennyson because his life was less pure, does Grey Owl's achievement lose much of its value ... what after all are we to make of an Englishman who denied blood and birthright, and of a husband who left to mourn his loss, as a dazed public was soon to learn, not one widow but four?

For many of Grey Owl's followers it was the awareness of being deceived, led astray, that made objective judgement of his work thereafter impossible ...

21. ... The name of Grey Owl will have different meanings for different people. For some it will bring memories of a tall, lean Englishman, Archie Belaney, who with moccasins, long hair and buckskin clothing, succeeded in passing without question as a North American Indian and winning international attention. To others, the name will suggest mainly a naturalist who projected his

beaver, McGinty and McGinnis, Jellyroll and Rawhide, to fame. And to still others the name will always recall the trapper who was converted to a genuine feeling for wildlife and became one of the most dedicated of conservationists ...

... In many ways, Grey Owl was years ahead of his time. At this period in Canadian history, conservation was not a popular topic. Were he living today, he would find many more sympathetic listeners when he talked about the wilderness and its denizens ...

22. ... What after all, does his ancestry matter? The essential facts about his life are not in dispute, as conservation officer under the Canadian government, and as lecturer and Broadcaster in Great Britain ... those who have read Grey Owl's books or heard his broadcasts cannot doubt his sincerity.

23. ... The camp is known as Beaver Lodge; yet to us it will always be a replica of that House, the empty Cabin, that is so far away – a little richer perhaps, a little better built, but the spirit is the same. Save for the radio, the kitchen range, the shape of the roof, it is as near to it as maybe. It is built of logs; the windows face out on the groves of trees. A painted warrior stands post in his appointed place, his eagle bonnet spread in brave array; the paintwork, the emblems, they have all been reproduced. The tokens are all there.

Atavistic? Perhaps it is; but good has come of it ...

KEY TO QUOTATIONS:

1 – Edward McCourt, in *Saskatchewan* (Canada Travel Series) (1968); 2 – Grey Owl, in *Pilgrims of the Wild* (1933); 3 – Hugh Eayrs, in a foreword to *POTW*; 4 – Grey Owl, in *The Men of the Last Frontier* (1931); 5 – Eayrs; 6 – Grey Owl; 7 – Eayrs; 8 – Anahareo, in *Grey Owl & I* (1972); 9 – Grey Owl; 10 – McCourt; 11 – Grey Owl; 12 – McCourt; 13 – Grey Owl; 14 – McCourt; 15 – Anahareo; 16 – McCourt; 17 – Grant McEwan, in a foreword to *Grey Owl & I*; 18 – Eayrs; 19 – Grey Owl, in *POTW*; 20 – McCourt; 21 – McEwan; 22 – Liverpool Daily Echo (April 21st 1938); 23 – Archie Belaney, in *Pilgrims of the Wild* (1933).

Agape:
Listening with Barry Lopez

Alec Finlay

If you go into the desert, silence no longer envelopes you. You become yourself such silence as makes the desert speak.
EDMOND JABES

There is a word from the time of the cathedrals: agape, an expression of intense spiritual affinity with the mystery that it is 'to be sharing life with other life'
BARRY LOPEZ

When I first visited Barry Lopez in his home at Finn Rock, Oregon on St. Valentine's Day 1993, the grouse flower was in bloom, a first flower of the spring; thinking back now to walks by the MacKenzie River and along the forest pathways at Finn Rock reminds me of Emily Dickinson's 'little Arctic flower/Upon the polar hem...'

Barry Lopez belongs in the tradition of American nature writing, of Cadwallader Colden, John Bartram, Peter Collinson, Mark Catesby, of Audubon, Thoreau, and John Muir of the High Sierras, and of modern poets, such as Robinson Jeffers and Gary Snyder. These writers chose their metaphors from nature, but they were often also moralists, whose concerns were with the individual, and with the ways that the life of the individual and the society he belongs to touch upon the wilderness. The hallmark of this tradition is an insistent inquiry into the relationship and responsibilities contained in that meeting place. When Nathaniel Hawthorne said of Thoreau, 'I think he means to live amongst us as an Indian', he may have been disapproving, but here was also an acknowledgment of the seriousness of Thoreau's sojourn, and of its precedents – recognition that, strange as his way of life might have seemed, here was someone attempting to live the life of a native New Englander. Choosing to set himself apart from the society of Concord, Thoreau sought a new way of life, one shaped by the woods. He understood that this would run parallel to the lives of the first peoples, those who had made a life amongst the lakes and the forests before him,

whose names and remains he studied with such keen interest.
In a similar way Barry Lopez seeks evidence of ways of living in the new world. Like Thoreau, he watches and listens before he writes, and learns from those who have gone before. Walking and looking are skills to be learnt. In this interview he describes the experiences of some of his sojourns, the word itself seeming to offer an invitation, 'so journey'; his experiences alone and in the company of indigenous communities, in the wilderness and the deserts of America, Australia, and in the Arctic.

Before the 'sojourn' in the wilderness became his primary subject Lopez wrote Of Wolves and Men, *a book in which he records how he learnt about nature from two semi-domesticated wolves he and his wife kept. Walking with these wolves was a way of attuning himself to the woods: 'by behaving as they did – minutely inspecting certain things, seeking vantage points, always sniffing at the air. I ... felt vigorous, charged with alertness.' These experiences, this learning to attend to what was near, helped him when he went to the Arctic. There he recognised how different was the world the Inuit saw. These journeys were, finally, an inquiry into his own life, his own ethics, his own emotions. As he journeyed his quest for knowledge, in all of its forms, was transformed: 'I sought, finally, companionship ... not reason, not explanation.'*

'Years ago this Inuit put his arm around my shoulder, and he said: 'You ... you are always thinking' – and he didn't mean it as a compliment! What he meant was that I was so conscientious – in studying the environment, in posing questions, in doing my research, in being careful with language, I was living too much in my mind. When I was travelling with people I should stop thinking and be *in* the place. That was one of the strongest lessons I learned from travelling with indigenous people, to not think about it, to be in it...

'The generalities all of us deal in are all right to have just to get down the road, but they're not specifically useful – and they can be dangerous. To think that you understand anything – to think that you understand an animal, or another human being, or a building – you can talk about it, more or less articulately, and you can relate it to one or another thing that occupies the same visual pattern, or the same biological pattern, or the same metaphysical pattern – but you can't really *know* that thing. I don't believe that's possible. Once you get outside of a self-referential world, you awake to that idea...

'We're so abstracted from what we call the natural world – always a bit dangerous, saying that, because it implies we're not natural – and that world is too often seen as an entertainment of one sort or another, or a type of distraction, or scenery. We've been taught to think that the small and insignificant *is* insignificant – that when we pay attention to nature, it should be when a volcano blows, or when there's flooding, or some other kind of so-called 'natural catastrophe' – where, in fact, the case is that *we* are a small part – humanity and its incredibly complex cultures – are a small part of nature. If you're not conscious of where you fit in this pattern then inevitably you'll get yourself in trouble...

'When I think of who my colleagues are as writers, I don't necessarily think about those who use the same metaphors I do. The metaphors I was born to, because of the way I grew up, were natural history, and geography, and anthropology, and archaeology. Those are the epistemologies, if you will, the ways of thinking about the world, that I felt comfortable with. When, after a very long period of time, I realised that I was a writer, that realisation came because I understood I was trying to write about 'important issues': justice, and tolerance, and dignity; but the way for me to get at those things was through these metaphors I felt comfortable with...

Much of Lopez's writing has a sense of 'fundamental anguish'. It echoes the vision of life characterised in Jeffers' poetry: 'It is needful to have night in one's body.../ The lonely-flowing waters, the secret keeping stones,/ the flowing sky.../ These things wash clean the mind.'

He has written four volumes of short stories, Desert Notes, River Notes, Winter Count, *and* Field Notes. *However, his epic prose essays,* Of Wolves and Men *and* Arctic Dreams, *remain his greatest achievement. In these two books there is a gradual movement, from writing about nature as a moralist – exploring the possibility of a 'just relationship' with the natural world – towards writing about nature which is more metaphysical in tenor. What Lopez has to say about human society and the problems that confront it is increasingly framed within a personal search for spiritual meaning.*

In Arctic Dreams *Lopez examines the lives of the Inuit and other first peoples of the Arctic. He also recounts the expeditions of European and American explorers – Peary, Franklin, Stefansson, Parry, and others – their tragic folly, their bravery, cunning, perseverance and their self-delusions. He records how these men were gradually turned from their intended purposes, exploration or exploitation, as the Arctic landscape imposed itself, creating a sense of: '... awe and consternation ... Its power flows into the mind from a realisation of how darkness and light are bound together within it, and the feeling that this is the floor of creation.'*

In the course of Lopez's own journey through the Arctic, light becomes his primary subject of inquiry, and his most enduring metaphor: an extreme physical experience, when he describes the fear brought on by the winter darkness as the mind 'draws into itself', or the exhiliration of summer's endless days. Gradually this physical experience of light becomes the locus for a sense of loneliness, loss, terror and awe – emotions that emerge in this severe and beautiful landscape, and which are illuminated, in retrospect, by his recourse to a metaphysical definition of love drawn from the age of cathedrals, 'agape'.

'When people ask me what I feel when I'm walking in a landscape, I say I'm looking for a way to be in love. I'm trying to open myself up and take advantage of any opening. What you are looking for in the woods is this elusive 'I – Thou" relationship of Martin Buber: 'I am this person, and I wonder who are you?' You begin to pose questions, as it were, to the land. I've sometimes called it a gentle interrogation, because I don't think it's right to go into the woods and pose a lot of questions. What you're doing is going out and opening yourself to the place, and once the place trusts you, it responds to you. The reason we feel made whole, or elevated, by that experience, is because we live alone. To not live alone is to be in love.

'When you go into a landscape and you feel that the landscape knows you are there, that it responds to your presence, then I think you have a sense of being less anxious in the world, more at ease about yourself, about everything. And it's in this dismantling of loneliness that I find a literature of hope. If I am in a very remote landscape with indigenous people, trying to understand animals that occupy that landscape through their eyes – if I come back and write about that, someone could say, well, this is a kind of 'travel writing', like Chatwin did, or someone else could say, well actually this is a kind of 'natural history writing', or 'nature writing', because of my interest in the biology and ecology of polar bears, or something like that, but what it is for me is a sojourn – often in the Arctic, or in the desert – in landscapes where the lines are classical. That's a visual environment in which I feel comfortable, more so than in a baroque environment, like a rain forest.

'It's in those places, by addressing very simple issues, that I can state, and restate, and then address again, the things that I'm interested in as a writer. Which would be, for example, what is the nature of a just relationship? If I could say one thing in my whole life as a writer, if I could leave some trace, it would be to help define – in a world rapidly changing – how we are to maintain just relationships. In recent years I've been thinking about what informs just relationships, and what must inform a just relationship is love.'

Barry Lopez lives with his wife Sandra by the McKenzie River in Oregon, on the western slopes of the Cascade Mountains: In Occupancy (1993), he describes how:

'In historic times, Molala and Kalapaya Indian people camped in this river valley when thimbleberries, red huckleberries, blackberries, elderberries and salmon berries were ripening and salmon were running ... Sandra and I arrived here in the spring of 1970 ... Our physical and spiritual connections to this landscape are myriad – plant roots, each with many rootlets. What has grown most slowly over ninety-four seasons here is an emotional attachment, an emergence of feelings akin to those one has for some members of a family or lovers or close friends ... For a while we are its human companions...'

On my most recent visit to Lopez's home, on the short walk from the cabin I was staying in up to his house, three logging trucks passed me on the highway. They were loaded with trees already tall when Columbus landed in America, five hundred years ago. Douglas Fir, Cedar, Hemlock. All clearcut.

'I live in a country – in the United States – where people are obsessed with being in power, not with being in love. There are many kinds of love – platonic love, spiritual love, romantic love, sexual love – there are many kinds of love, but they have some things, to me, in common. One is intimacy of one sort or another, and one is vulnerability...

'In order to love someone you have to accept their deep and unchangeable flaws. You can't just say, well, I'll love this part of you, and the other part I'll ignore. I think, for myself, and for other writers, the impulse to write is very close to the impulse to love. As I've grown older what God has come to mean for me – that which I call God – has become apparent in the most intense acts of loving – loving relationships...

'I remember writing something in *Arctic Dreams* that surprised me when I saw it on the page, but when I looked at it later I realised it was something I had wanted to have in front of me for a while. I had been talking about a kind of love called 'agape', the love of God in another person, seeing God in another person and loving the person because of that manifestation. I had talked about other kinds of love, and alluded to the difficulties that humanity finds itself in, historically ... Tamberlaine, the Black Death, situations where we're prone to despair, but then somehow we get out of it. We often believe that we get out of these diffIculties because we can think, that it's our intelligence that always saves us. What I wrote was that this is true, but we don't know whether intelligence is reason, or whether intelligence is love. In the years since I saw these words on a piece of paper in front of me, I've thought I want to know more about this. What does that mean, intelligence is love? I think that it's possible to say that it's more intelligent to be in love than to be in power, or calculating how one might make one's life better by subjugating other people.

'There was a man, a mountain-lion biologist – I was working on a story about mountain-lions and I went down to Arizona to stay with him for a while, to look for lions and talk – and we were out on the rim of the Grand Canyon very early one morning. The sun had just come up, and because its rays are so low it's easy at that time, very early in the morning, to see the shadow and light in lion tracks. So we were driving along very slowly in the pick-up truck watching for lion tracks in the dust in this road, and in that silence he said to me: 'It's not in the data Barry'. And I said: 'What do you mean?'. He said: 'What I know about mountain lions is not in the data, it's in here' – and he made a gesture to cover his face and his head and his heart, and I understood what he meant...

'I understood in the same moment the frustration of the best biologists I'd been with in the field; which is that so little of what they really know finds its way into scientific papers, because the journals restrict the expression of information, and scientific credibility counts for more than full disclosure...

'So I tried to put him at ease, because he was an acquaintance and I liked him, and I said: "Well, you know, we depend on you as a scientist; the scientific method, for all its flaws, it's a good way of knowing. If you give us the data perhaps that's the only way we're ever going to be able to do anything about protecting the mountain lion in northern Arizona, and coming to an understanding of that ecosystem. We depend on you to do this as a scientist..."

'He replied: 'But we're not scientists, we're historians'. This was like the proverbial lightning bolt when he said it, because I understood in that moment what I knew so many wildlife biologists to feel, which is that they don't *understand*. You don't come to an understanding of a species, you come to an understanding of the interaction of an individual animal over time with place. The way he put it to me was, after he had studied mountain-lions for a year he would be willing to generalise about mountain lion behaviour for lions in Florida or in northern British Columbia or in Patagonia, wherever they live; that it's pretty much the same everywhere. But now, after ten years, all he was willing to say was: 'If you wish me to discuss mountain lions, I'm willing to talk about these mountain lions here, on the north rim of the Grand Canyon, over the past ten years, but what other mountain lions might do I don't know...'

'Western people often have a desire to get a quick summary together. Somebody will say, what do wolves do? Or polar bears? What does an animal do? Indigenous people will say, this animal at this time in this place did this. What they're saying is the richness of life consists in the wide range of expression – not in the reduction of that expression to the one or two pieces that fit into the industrial machine. It's why fifteen or twenty people can walk through a similar landscape and each come up with a different but valid story – a scientist might come away with a story about the behaviour of one or another sort of bird, and someone whose metaphors are urban might come away with a story about the tenderness of the relationship between a man and a woman who'd been estranged for a long time, who've come back together because he's spent the afternoon watching the behaviour of the same wind with leaves of different sorts. All of those stories tell us, again, that there is no one relationship between human beings and the world, any more than there's one relationship between polar bears and the ice...

'*Isumataq*' is the Inuktitut word for storyteller. It means *that person who creates the atmosphere in which wisdom reveals itself*: That's how they see the storyteller. I think that's very much how I see my own work: not as somebody who's wise, or has prescriptions, but someone who recreates a pattern, or an environment, from which wisdom will emerge – creating a set of relationships that make wisdom apparent. I do it with the knowledge that there will be people much brighter than I am who will read the story and see something more deeply.'

Alec Finlay orchestrates Morning Star Publications. His most recent publication is *Carmichael's Book*, 'a homage to the Carmina Gaedelica' featuring the work of a number of Scottish poets.

For Tom and Laurie Clark

Travelling Back Through Stories[1]

Allice Legat

THIS IS A CIRCULAR STORY. It is about how my grandpa's stories led me to my work with the Taku River Tlingit of northern British Columbia, Canada and the Dogrib of the Northwest Territories, Canada.[2] In turn their oral narratives led me back to 'where I am from',[3] which is Smerrel, Latheronwheel, Caithness. When my grandfather left Scotland in the early part of this century he had stories, both mythological and real. These provided him with a character and an identity. The knowledge contained in them gave him a relationship to the land and the ancestral spirits. His stories provided him with an opportunity to share his knowledge of these relationships as well as his knowledge of his Scottish character and political struggles. This knowledge did not, however, give him a wage with which to support his family. He left Caithness, Scotland and travelled to Saskatchewan, Canada, where he settled and supported a family of eight as a blacksmith and farmer.

When I travelled to Scotland in 1995, I too had stories, also mythological and real. Some of these provided me with an identity that evolved out of the vague memories of my Grandpa's. Despite my ancestry, more immediate to my identity as a Canadian were the Taku River Tlingit and Dogrib stories. These provided me with both the knowledge that I live on native[4] land, and an understanding of the struggle aboriginal people have to maintain their character and a relationship to their ancestral spirits. Through these stories I began to see the Scottish struggle within the British state, and the native struggle living on crown land in Canada, as similar.

BEGINNING LIFE WITH STORIES

I first heard oral narratives from my grandfather. As a very small child I remember being lifted up and placed on the table to listen and watch. To listen to my grandpa tell and sing; to listen to my grandpa play the fiddle; and to watch my mom dance to Scottish tunes. These were happy times. I do not remember the names of the tunes or the dances. I remember the sound of the fiddle, and I remember the visions that came to me as I listened: boys climbing the cliffs to retrieve seagull eggs; of the crofts by the sea; of the sheep dogs; of the rain and fog; of the Clearances; of his dislike of the English, church and liquor; of herring fishing; of brothers, cousins and in-laws dying on fishing vessels; of women carrying baskets of herring to the market in Wick; of his struggle to find work in Glasgow; and of the origins of the clans. He also told us mythic stories of fairies and witches, and of giants who were buried within hugh cairns on the Caithness landscape.

Grandpa Campbell died when I was five. After he died our home was a little quieter. My mom tried to keep the Scottish character alive in us. For her, to have Scottish ancestry was the greatest gift any human being could have. She reminded us of the Highlands, of Caithness, and she periodically spoke a few phrases in Gaelic. She talked of my grandmother who stressed the importance of education and integrity, and that it was the Scottish who first had schooling for all. She would tell how grandma would say, 'No one can take your knowledge; they can take your land and your home, but they can never take your knowledge.'

Mom would talk of their life in Canada. She would tell stories about how 'Indians' would come to have their horses shod, and when they offered to pay, Grandpa would say, 'Haven't we taken enough, we have your land.' She would also tell how grandma (a Hendry) told her of the shame the Campbells had once brought to Scotland, and when hearing this grandpa would go to his shop. As I sat and listened I thought about my Grandpa Campbell, and wondered how the Campbells could be so bad if my grandpa was from that clan.

Although my mom tried to keep the Scottish character alive, there were no more stories, songs, fiddle tunes like his. He had walked the land in Caithness, not her. He felt the love and passion in his people, she felt his love. As I grew to be a teenager I began to forget the stories of Scotland; the stories of our ancestral past. But they lingered inside of me, moving me in a way that caused me 'to live my life like a story'. I travelled north towards the Canadian Aurora Borealis. In following the northern lights I met Taku River Tlingit and Dogrib elders who stirred my memory and I came closer again to the stories of my ancestors.

LIVING A STORY ON TAKU TLINGIT LAND

It was the elder, Elizabeth Nyman, a Taku River Tlingit, who taught me to love oral narratives again. She reminded me that it is stories that give us our character and identity. She made me walk the land as she shared her oral traditions. She awakened in me those things that you can only know about yourself if the stories are allowed to reach your heart. She did this by telling me about her moiety[5] and clan history; about the hunting and finding of moose, sheep, caribou, seals, goats; about living on the land; about the first 'whites'[6] from Russia; and about the thousands of 'whites' who arrived on Tlingit land during the 1898 gold rush.

I could visualize, while listening to her, 'whites' through time searching the Canadian North for resources,[7] building houses but never really settling. Perhaps because 'our roots' are in stories connected to other ancestral homes, and unless we listen to the stories of the aboriginal people we are unable to connect to our immediate landscape. I wondered if this happened to the early invaders of Scotland. I have watched as 'whites' invade native land, ripping the resources from it while destroying burial sites, yet protecting burial sites in Europe, and Euro-Canadian settlers', homesteaders' and stakers' homes on Crown land. Although I knew of the Clearances and the separation of people from land and home, I wondered if the British state continued to take Scottish resources while destroying land and place, both mythological and real.

Hearing the Tlingit stories about the first 'whites' and the various encounters with the Russians and U.S.'ers, made me want to hear about first encounters between the Scots and the Picts, between the Vikings and the Picts. I wanted to hear the stories about the various encounters between the Scottish people and the British bureaucracy. The Tlingit have clans, and hearing their clan's stories created curiosity in me about my own ancestral clans. Hearing how the Taku River Tlingit came inland, made me want to hear the oral traditions of the Campbells of Argyll, and how my ancestors settled in Caithness. I wanted to hear my people's stories.

TRAVELLING BACK THROUGH DOGRIB STORIES

It was thirty Dogrib elders who taught me to see the importance of stories in maintaining idealism in character and behaviour, which is important to peace, and respect of others' place and their connection to their land. They told their cosmology and history starting with creation and mythological time, and ending with the present. They told of Federal laws being imposed without consulting them; they told of 'whites' who took over their land without ever walking on it.

They shared stories about their land, the land I live on; they shared the spirit that has always governed the northern landscape. The Dogrib elders told me stories about the similarities between the

Scots that came to northern Canada and the Dene.[8] The elders would make statements such as, 'The Scotsmen who came here were like the Dene, they danced, laughed and lived close to the land.' Their stories connected me to my past and gave me an understanding of a Scottish attitude through the eyes of people who had watched 'whites' come and go.

And I wanted to see what my grandpa had told me about his life, about the Caithness landscape, and about the people. I wanted to understand this Scottish character that lived close to the land. I wanted to see more closely the people with whom the Dogrib had intermarried. I wanted to discover if there is a Scottish attitude of the mind, and if it is the base for my intuitive passion about the taking of people's land and the denial of self-government. I wanted to know through oral tradition why my grandpa had seen so clearly that he lived on native land. I started to wonder how the Scottish stories and songs reflected the importance of land, place and governance. I wanted to hear crofters' stories.

The Dogrib elders are focussed. They are constantly thinking about how to tell their grandchildren about their ancestors, about the ancestors' knowledge. They are concerned that young people maintain Dogrib character, yet learn new technologies and ways of thinking. I wondered if there are crofters' stories that tell about technologies and skills that were learned from Norwegian, French and Dutch.

To the Dogrib both schooling and the elders' knowledge is important, so the young people will be 'strong like two people'.[9] Watching the Dogrib elders strategize so their knowledge will become part of the school curriculum caused me to wonder at the state of Gaelic, both in Scotland and in the Maritimes of Canada. I began to understand how my grandparents must have felt as they watched their children learn only English; losing Gaelic and, therefore, the link with Gaelic stories, ancestors and homeland.

The Dogrib elders think of their traditional territory as the home for their grandchildren. They continually tell them of the land, their home. Like my grandpa they never give up telling their grandchildren that the land is theirs not the Canadian governments – that both the English and the French are new-comers (500 years) and colonizers.

Hearing Dogrib origin stories made me wonder about the origin of my ancestors' clans: the Campbells, the Gunns, the Sinclairs, the Hendries. Hearing the stories about traditional Dogrib governance created a desire in me; I wanted to hear, feel and think about the clan stories, and the bigger Scottish story; the stories being brought forward from the ancestors; the stories that take a person back through time, connecting person to person. I wanted to hear about the footsteps each person took.

I still only know a little of the place where I have chosen to raise my son, and the Dogrib, with whom I work. I listen to the stories of the Dogrib elders. Stories that tell of Dogrib origins, wars, philosophies, and the first 'whites'. I started to wonder what Lybster looked like; what about those cliffs?

TRAVELLING THE STORY ON SCOTTISH LAND

My son and I travelled back to Caithness, Scotland in 1995. The stories, coupled with our journey, caused me to think of both the Dogrib and the Scottish as political intellects; as people who use oral tradition to remember their character and identity, and to remember the land and what it is they need to maintain their character and identity. I walked the stories I had heard from my grandfather. I saw the origins of the stories that initially directed my life.

As my son and I travelled back, I understood more clearly the similarities between the native people in Canada living on Crown land, and the Scottish people within the British state. I understood the Dogrib elders' stories that reflected how Scots were like them, and my mom's stories about my grandpa's refusal to take money from the native people in Saskatchewan. Land and self-governance is paramount to both. Through these stories I also began to understand the importance of the relationship between the crofters, and the land and the issue of ownership, within Scotland.

I began to question why so many people in both Canada and Scotland compare the desire for Scottish independence from England with the issue of sovereignty in Quebec. Based on the oral narratives of the people, I think the political issues faced by the Scottish and those faced by French Canadians have little in common. Both the French and English wanted to control aboriginal land and resources, just as the English intention was always to be the dominant partner in union with Scotland. I began to think of the aboriginal situation and political struggle in Canada and the Scottish political struggle in Great Britain as similar: for over 300 years both have tried to re-claim their right to self-government and their right to govern their own lands and resources. The Gael and Scot, Dogrib and the Taku River Tlingit use oral traditions to remember traditional land and ways of governing; these stories are passed between people and through generations so the young remember that they have the right and the will to be where they are and to govern themselves.

As we travelled north towards Caithness we honoured the connection between the Gaelic and Dogrib speakers, and both people's connection with their land. We listened to the people. In Glasgow we remembered my grandpa's stories of the workers. We walked, as working class through the castle gardens of the Campbells of Argyll, and felt very separate from the contemporary owners. As we drove into Caithness and heard the fiddle, we explored the cliffs where seagull eggs were gathered and saw the crofts where children and animals had once been secured so they would not fall over the edge. We heard more stories of the Clearances, and how the people were forced to live on these dangerous cliffs. We saw Campbell graves in the churchyard[10] at Latheronwheel. We met relatives who were holders of oral tradition and who told stories. We visited the place where women had climbed the hundreds of stairs with their baskets of wet herring as they walked to the market in Wick.

We heard about the 'white settlers'[11] who are currently paying more for the land than most crofters can afford. We also heard about the crofters' anger at losing both what they love and the essence of who they are. I wondered if this could be the final clearance? We watched the tour guide point to the old Royal High School in Edinburgh, as she proudly explained that this would house a Scottish Parliament once self-government is re-claimed.

I understood more clearly why I am so passionate about the taking of people's land and the denial of self-government. All the elders I have known have instilled in me the pain people feel when they lose house, hearth, land and language.

On the Orkney Islands we saw the schools that were built with Hudson's Bay Company money. We appreciated the importance of education in Scotland, yet began to see how the Scottish working class had been used. Oral tradition tells of men being sent to Canada with one way tickets and how they were encouraged to become trading post clerks, thus becoming incorporated into 'The Company'. The Hudson's Bay made money from the furs trapped by native people. I wondered how many native children were educated with Hudson's Bay money, as I know of no schools built by them for native children.

In Caithness, we heard of people trying to learn Gaelic, just as the Taku River Tlingit are trying to learn Tlingit.[12] We also heard the origin stories of the Gunns and the Sinclairs, and saw the site of the battle that had brought the Campbells to Caithness to fight the Sinclairs.

We walked the land and listened to stories as we visited. We sat on the ground watching the sheep with a 94 year old elder, a member of the Campbell clan who looks a lot like my grandpa. He stated, 'The trouble with young people today is they do not listen to the animals.' I thought of the 94 year old Dogrib woman who has often said, 'If they really want to know about the land they should ask the animals.'

We tried to honour the relationship between the Scottish people who came to Canada and their relationship with aboriginal people, especially those who had married Dogrib. We also tried to honour the relationship we felt between the prairies, where my grandpa settled, northern Canada where I settled, and Caithness, where our ancestors are buried. We thought about the Dene drum when we heard the bagpipes; we thought about the caribou while we watched the sheep and the Highland cattle. We thought about the unmarked burial sites on Dogrib traditional territory as we heard the story about the buried giant who would come to life if the huge cairns were disturbed. I wondered what will protect the ancient burial sites on Dogrib traditional territory as the several diamond mines are developed. What if the mine management does not listen to the stories – both mythological and real? I could not help but think that although we have heritage legislation, few sites are documented. I wondered which of the diamond miners will listen to the Dogrib elders' oral narratives and protect the ancestral spirits?

In Caithness, we heard an English woman defend her purchase of a heritage house. She explained that the Scottish are a 'lazy race', and that if she did not have it the house would probably not be cared for, and that if resources, such as oil, were not being developed, there would be no infrastructure for the British state to support the Scottish health and education systems. I wondered if she would ever realize that Scottish resources such as oil, whisky, sheep, are an important tax source for the British state. I couldn't help but compare this with the number of times I have heard 'whites' in Canada call 'Indians' lazy and make the argument that it is important for the government to develop resources on native land so there is an infrastructure to support and employ aboriginal people.

Oral narratives continue to describe 'whites' and 'white settlers' who do not hear the stories told by Dogrib and Taku River Tlingit elders and the Scottish crofters. Is it possible for Canadian and British parliaments to live the stories that they have been told for centuries; the stories that tell of people governing themselves on their traditional lands? It seems that some Scottish politicians can.

By mid-September, 1997 the Scottish people will know the results of a referendum on devolution, and by fall, 1998 the Dogrib, who are among the first aboriginal people in Canada to negotiate self-government in their land claim, should know their relationship with the Federal and Territorial governments. During the discussions leading up to the referendum, I was once again struck by the similarities between the Scots and the Dogrib when I read:

> ...there is surely everything to be said for the kind of politics Donald Dewar so powerfully and poignantly represented in London and Edinburgh last week; a politics of decency fought for with a passion, a politics that prefers consensus to conflict, compromise to intransigence, and the long hard search for true mutual respect and real common ground, over the barren assertion of dominance on one hand, and irreconcilable difference on the other.[13]

This statement reminds me of several very similar comments made by Dogrib elders who search for mutual respect and harmony with the Federal and Territorial land claim negotiators. Statements[14] such as: 'we do not want to embarrass them; we want the 'whites' to know whose land we are discussing; we do not want financial compensation, we want the Federal government to understand whose children played on the uranium tailings in the 1950s[15], and whose grandchildren will be here living on the land after the diamond mines are closed.'

THE CIRCULAR STORY CONTINUES

I am back on Dogrib traditional territory. At times I listen to oral narratives on placenames, habitat, governance, and land. My identity is a little stronger now, and I know the Scottish attitude has a lot to do with my intuitive passion about the taking of people's land and the right to self-government. I can see this attitude and passion in my son. At other times I tell my mother of my experiences in Scotland and with the Dogrib – stories that are both mythological and real.

NOTES

1. I first heard of the concept of living one's life like a story from the Tutchone and Tagish in the Yukon territories, Canada. Julie Cruikshank's *Life Lived Like a Story* (1990), published by the University of Nebraska Press contains the stories and lives of three Yukon Elders. The Dogrib elders make similar statements.
2. The Taku River Tlingit traditionally have permanent long houses on the northwestern coast of Alaska. The Taku River Tlingit used the Taku River as their travelling route into the interior of northern British Columbia and southeastern Yukon. The headwaters of the Taku River are located in the coastal range of northern British Columbia and empties into the Pacific Ocean near Juneau, Alaska. Inland they traded with the Tutchone and hunted goats, moose, inland caribou and sheep.

To protect their inland territory, some clans made more permanent villages around Teslin, Yukon Territory and Atlin, British Columbia. The traditional lands of the Dogrib are located between Great Slave Lake in the south and Great Bear Lake in the north, and between the Mackenzie River in the west and into the barren lands to the east and northeast.
3. The old people in Caithness would ask, 'Where are you from?' I would answer, 'I am from Canada.' There would say, 'No where are you from?' I finally realised they meant where do I trace ancestry.
4. 'Native' and 'aboriginal', are used interchangeably here as they are the preferred terms used by the Dogrib and Tlingit when speaking in generalities to non-native people.
5. 'Moiety' refers to a primary social division in that the Tlingit are made up of two exogamous groups, each of which contains several clans.
6. 'White' here is used as northern native peoples use this term. This can include anyone, regardless of colour, who belongs to the dominant society and is perceived as trying to control native land, and people such as bureaucrats and individuals working for transnational corporations. Individuals can be seen as 'white' in one situation and not in another.
7. Throughout the Canadian north resources such as furs, gold, uranium, oil, lead, zinc, and now diamonds have been coveted and taken by transnational companies.
8. 'Dene' means 'the people'.
9. This direction was given by Chief Jimmy Bureau in the 1960s. Jimmy Bureau was the last traditional chief, according to the elders. He told his people to always know their own knowledge and ways, as well as the knowledge and ways of the dominant society.
10. One section is referred to as Campbell's Corner and contains several unmarked graves. Oral tradition tells that these people criticized the Church of Scotland and were therefore refused the right to be buried in the churchyard. Since it was not possible to be buried elsewhere, a gravedigger secretly buried them in the corner and kept records for the family.
11. The term 'white settler' seems to be similar to 'white' in that neither really refers to colour, but rather to being from elsewhere and having an affiliation with those who are disrupting an established way of life. In Caithness I asked if I would be considered a 'white settler' if I came to live there, and they said, 'No, you are from here.'
12. Dogrib children enter school speaking Dogrib as a first language. The Dogrib are attempting to develop a curriculum and other programs so their language will be maintained and continue to develop.
13. McMillan, Joyce, July 27, 1997. *Scotland on Sunday*, p. 15.
14. This is not a verbatim quote, rather a conglomerate of several statements I have heard from Dogrib leaders.
15. Dogrib Renewable Resources Committee, 1997. *The Trees All Changed To Wood*, Rae, NWT, Canada.

'...while the missionaries prayed...'
extracts from her autobiography

Emily Carr

A MISSIONARY TOOK A LIKING TO ME. She had a long face but a good heart. She was negotiating for my sister to accompany her back to her lonely mission up the west coast of Vancouver Island, so that she might try out the loneliness and the Indians. When the missionary saw how interested I was in her descriptions of these wild places, she said to me 'Wouldn't you like to come to Ucluelet to sketch in the summer holiday?'

The *Willapa* was a small coast steamer. I was the only woman aboard, indeed the only passenger. We nosed into dark little coves to dump goods at canneries. We stood off rocky bluffs, hooting until a tiny speck would separate itself from the dark of the shoreline. It grew and presently sprouted legs that crawled across the water. The black knob in the middle was a man. We threw him a rope and he held on, his eyes chewing the parcel in the purser's hands, his face alight.

'Money?' shouted the purser. The man's face unlit. He made a pretence of searching through his ragged clothes and shook his head. The purser threw the parcel back on our deck and tossed a letter into the man's boat. The man ripped the envelope, tore out his remittance and waved it, the parcel thudded into the boat! We tooted and were away. Then it was gone. Vastness had swallowed the boat and man ...

... Life in the Mission House was stark, almost awesome, but you could not awe our missionary, she had no nerves. She was of cement hardened into a mould. She was not inhuman, there was earth underneath. It was just her crust that was hard and smooth. The slow, heavy Indians had not decided whether or not to accept religion. They accepted missionary 'magic' in the shape of castor oil and Epsom Salts. But religion? They were pondering. The missionaries were obliged to restrain their physic-giving. If you gave an Indian a bottle of medicine he drank it all down at once and died or not according to his constitution. He had to be given only one dose at a time. But the missionaries expected to give the Indians the whole of religion at one go. The Indians held back. If physic was given in prolonged doses, why not religion?

'Toxis', as the Indians called the Mission House, squatted back to forest face to sea just above the frill of foam that said, 'No further,' to the sea and, 'So far,' to the land. The Indian village was a mile distant on one side of the Mission House, the cannery store a mile on the other. At high tide we went to them by canoe, at low tide we walked in and out among the drift logs lying stranded on the beach.

No part of living was normal. We lived on fish and fresh air. We sat on things not meant for sitting on, ate out of vessels not meant to hold food, slept on hardness that bruised us; but the lovely wild vastness did something to it all. I loved every bit of it – no boundaries, no beginning, no end, one continual shove of growing – edge of land meeting edge of water, with just a ribbon of sand between. Sometimes the ribbon was smooth, sometimes fussed with foam. Trouble was only on the edges; both sea and forest in their depths were calm and still. Virgin soil, clean sea, pure air, vastness by day, still deeper vastness in the dark when beginning and endings joined.

After the missionaries blew out their candles and the ceiling blackened down to our noses, the square of the window which the candle had made black against the outside dark cleared to luminous greys, folding away mystery upon mystery. Out there tree boles pillared the forest's roof, and streaked the unfathomable forest like gigantic rain streaks pouring; the surge of growth from the forest floor boiled up to meet it. I peered at it through the uncurtained window while the missionaries prayed.

To attempt to paint the Western forest did not occur to me. Hadn't those Paris artists said it was unpaintable? No artist that I knew, no Art School had taught Art this size. I would have to go to London or to Paris to learn to paint. Still the French painters who had been there said, 'Western Canada is unpaintable ...'

... The first year that I lived and taught in Vancouver my sister Alice and I took a pleasure trip to to Alaska. The Klondyke rush had been over just a few years. We travelled on a Canadian boat as far as Skagway, end of sea travel for the Klondyke. Prospectors had left steamers here and gone the rest of the way on foot over a very rough trail. Those who could afford to took pack beasts; those who could not, packed their things on their own backs.

The mushroom town of Skagway had sprung up almost overnight. It consisted of haphazard shanties, spilled over half-cleared land. The settlement lay in a valley so narrow it was little more than a ravine heading the shallow, muddy inlet. Three tipsy plank walks on crooked piles hobbled across the mud to meet water deep enough to dock steamboats. Each runway ended in a blob of wharf.

Jumping gaps where planks had broken away, I went out on to the end of one wharf to look back to Skagway. Bits of my clothing and sketching equiment blew into the sea.

Wind always moaned and cried down the valley, smacking this, overturning that shanty. The little town was strewn with collapsed buildings. Crying, crying, the valley was always crying. There was always rain or mist or moaning wind. It had seen such sadness, the valley – high hopes levelled like the jerry-built shanties, broken, crazed men, drinking away their disappointments.

A curious little two-coach train ran up the valley twice a week. Its destination was White House. We took it up as far as the summit. Here, side by side, where their land met, fluttered the two flags, British and American. The train jogged and bumped a good deal. We passed through valleys full of silence, mocking our noisy little train with echoes because our black smoke dirtied the wreaths of white mist which we met in the valleys.

A stark brooding surly land was this, gripping deep its secrets. Huddles of bones bleached by the wayside – bullock bones, goat and horse bones, beasts for whom the lure of gold did not exist but who, broken under burdens of man's greed, had fallen by the way, while, spurred on by the gold glitter, he shouldered the burden of the dead beasts and pushed further. In the valley, drowned under lush growth, we stumbled upon little desolate log shanties, half-built. The finger of the wild now claimed the cabins, saying, 'Mine, mine!', mossing them, growing trees through the mud floors and broken roofs.

'How do these come to be here?' I asked.

'Failures – men who lacked the courage to go home beaten. Most drank themselves to death. For many a gun was their last and perhaps their best friend.'

No wonder the valley cried so! Of those who stayed in Skagway it was seldom said, 'He made good. He struck lucky!' The lucky ones went down to the cities, scarred by their Klondyke experience, seldom happier for their added gold.

We stayed one week in Skagway; then, taking an American steamer, crossed to Sitka on Baronoff Island.

Sitka had an American army barracks, a large Indian village, and an ancient Russian church. It also had many, many very large black ravens, sedate birds, but comical. They scavenged the village, calling to one another back in the woods. The male and the female made different squawks; the cry of one was a throaty 'Qua', the other answered, 'Phing!', as sharp and strident as a twanged string. At the top of its flagpole the barracks had a gilt ball which held a great attraction for the ravens. There were always three or four of them earnestly trying to get firm foothold on the rounded surface. It was amusing to watch them.

As we walked through the little town of Sitka I saw on a door, 'Picture Exhibition – Walk In'. We did so.

The studio was that of an American artist who summered in Sitka and wintered in New York where he sold his summer sketches, drab little scenes which might have been painted in any place in the world. He did occasionally stick in a totem pole but only ornamentally as a cook sticks a cherry on top of a cake.

The Indian totem pole is not easy to draw. Some of them are very high, they are elaborately carved, deep symbolicical carving, as much or more attention paid to the attributes of the creature as to its form. The Indian used distortion, sometimes to fill spaces, but mostly for more powerful expression than would have been possible had he depicted actualities – gaining strength, weight and power by accentuation. The totem figures represented supernatural as well as natural beings, mythological monsters, the human and animal figures making 'strong talk', bragging of their real or imagined exploits. Totems were less valued for their workmanship than for their 'talk'. The Indian's language was unwritten: his family's history was handed down by means of carvings and totemic emblems painted on his things. Some totems were personal, some belonged to the whole clan ...

... We passed many Indian villages on our way down the coast. The Indian people and their Art touched me deeply ... By the time we reached home my mind was made up. I was going to paint totem poles in their own village settings, as complete a collection of them as I could. With this objective I again went up north next summer and each successive summer during the time I taught in Vancouver. The best material lay off the beaten track. To reach the villages was difficult and accommodation a severe problem. I slept in tents, roadmaker's toolsheds, in missions, and in Indian houses. I travelled in anything that floated on water or crawled over land. I was always accompanied by my big sheep dog ...

... Indian Art broadened my seeing, loosened the formal tightness I had learned in English schools. Its bigness and stark reality baffled my white man's understanding. I was as Canadian-born as the

Indian but behind me were Old World heredity and ancestry as well as Canadian environment. The new west called me, but my Old World heredity ... pulled me back. I had been schooled to see outsides only, not struggle to pierce.

The Indian caught first at the inner intensity of his subject, worked outward to the surfaces. This spiritual conception he buried deep in the wood he was about to carve. Then – chip! chip! his crude tools released the symbols that were to clothe his thought ... The lean, neat Indian hand carved what the Indian mind comprehended. Indian Art taught me directness and quick, precise decision ...

from 'Growing Pains': the autobiography of Emily Carr (Toronto, 1946)

Emily Carr (1871-1945) was born in Victoria, British Columbia. Her father's favourite, she studied art in San Francisco from 1889 until 1895, and in England, where she lived for a time among the St. Ives colony, until illness forced a return to Canada. Back in Victoria, she worked for as an illustrator for the local newspaper and, disastrously, as tutor to the the 'Ladies Art Club'.

During this period, Carr's interest in Native American culture developed and she returned to Europe, to France, where she again fell ill. After convalescence in Sweden, she studied water-colour technique in Brittany with the New Zealander Frances Hodgson, whom she met at the *Academie Colorassa* in Paris. Carr's paintings from this time show a move away from naturalism, towards Fauvism, and some of her work was hung in the avant-garde 'Salon d'automne' in 1911. Carr returned to Canada in 1912, where she set about fusing expressive modernist style with Canadian imagery, as in 'Totem by Ghost Rock' (1912), for a show in Vancouver. But the response to this work was entirely negative. She returned to Victoria, where she kept a boarding house during a period of withdrawal which ended only in 1927, when her work was shown in the east.

The turnabout in Emily Carr's fortune was due to the intervention of the Canadian ethnologist and folklorist Marius Barbeau, who had sought her out in 1915 and again in 1921, after hearing stories from his native American friends of a strange white woman who visited them, making pictures. It was on Barbeau's recommendation that the National Gallery of Canada contacted Carr, and over twenty of her paintings were shown to great acclaim in Ottawa alongside that of the 'Group of Seven', whose emphasis on Canadian cultural identity found a powerful precursor in Carr's painting.

Encouraged by this success, Carr returned to Victoria to paint, travelling still deeper into British Columbia'a forests. Her paintings became more intense, darker and larger in compass. From 1930 onwards, her subject was increasingly the forest itself – vast, curvilinear forms that seem ecstatically alive and all-enveloping. Her public success reached a highpoint during the 1930s, but by this time poor health meant that she was increasingly unable to paint. She turned instead to literature, winning the Governor General's award for *Klee Wyck* in 1942, the first of five books she published, mixing factual narrative with a poet's attention to the subtleties of language. On her death in 1945, a major memorial exhibition was organised in Ottawa, and in 1972, when a retrospective brought her work to the attention of a country debating independence, appreciation of her work reached new heights.

Gradually Going Magadan

Ian McKeever

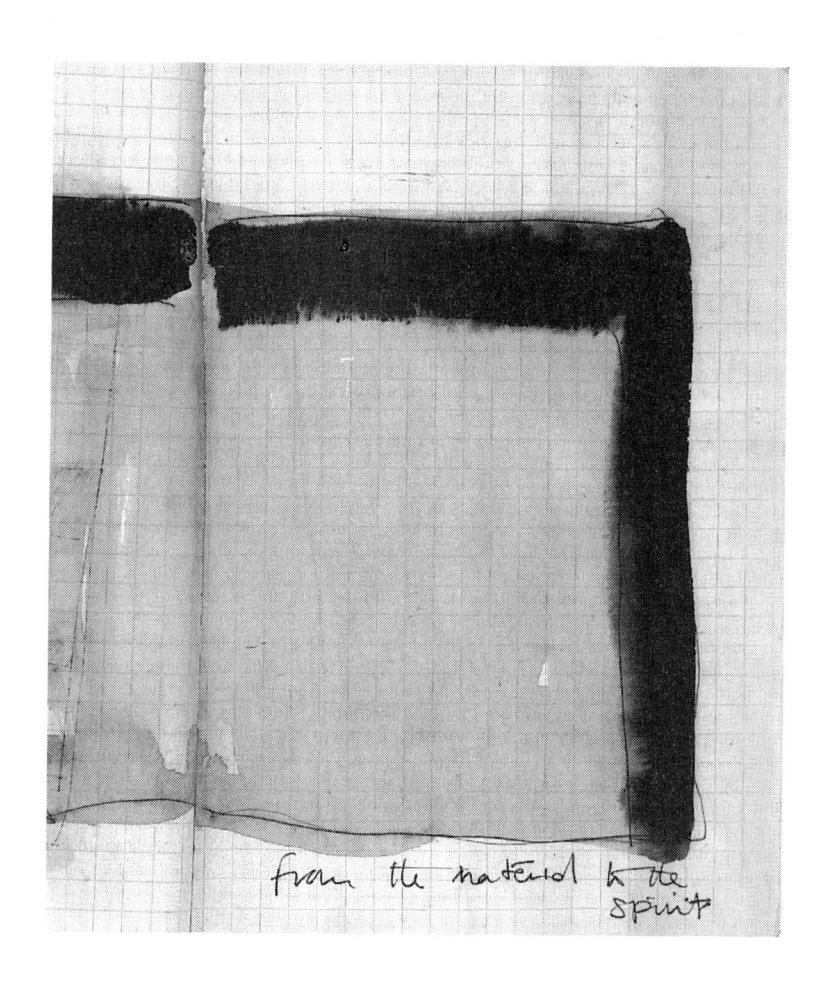

from the material to the spirit

*

I AM BACK IN MAGADAN. Again sitting in the kitchen of Berman's apartment. Berman is a scientist working at the Institute for Northern Research in Magadan. The last time I was here in the summer of 92, we travelled with a group to the tundra of the northern coastal plains; Pevek and the Chaun estuary. Then amidst scientists, watching them at work, I was struck by the similarities of how they go about their work and how an artist might work. Both seem to spend a great deal of time just looking and thinking around things. Engaged in the refreshingly simple activity of observation. Walking the line between passivity and reception where, if one is really lucky, insight may occur.

This time I am stuck in Magadan. Plans to travel on to Kamchatka peninsula stalled by the weather. No flights and no one seeming to have an idea as to when there might be one. Magadan under the snow is a different place to when I last saw it. The snow suits it. Somehow its crude concrete apartment blocks begin to become a kind of architecture, as they sit up square out of their footings of white.

So there is a lot of hanging around and killing time, but that is part, and often a good part of travelling. For one travels to be disarmed, and sometimes in those dead hours, which either have their own time, or no time at all, there is the possibility to just gaze and be open. To register those small details which make up the differences. For there is an edge where cultures meet and there is a second edge where the edges of cultures meet. This second edge – of often fleeting and disconnected details – is sometimes more poignant and disarming. Moments can then appear to sneak around outside of time.

It is early one evening, we are going to visit a local artist. Walking there with that special shuffle of keeping the feet in contact with the icy ground, in order to stay on one's feet. And the dull dark early evening light against the snow. Everything rounded and vaguely silhouetted, just as in a painting by Munch. A few months before I had seen his paintings in the museum in Oslo, and in the room next door a group of early ikons, and was struck by their connection. Thinking then as now of how his figures with their closed rounded half-silhouettes become forms as the flat light of snow night freezes them into ghosting ikons.

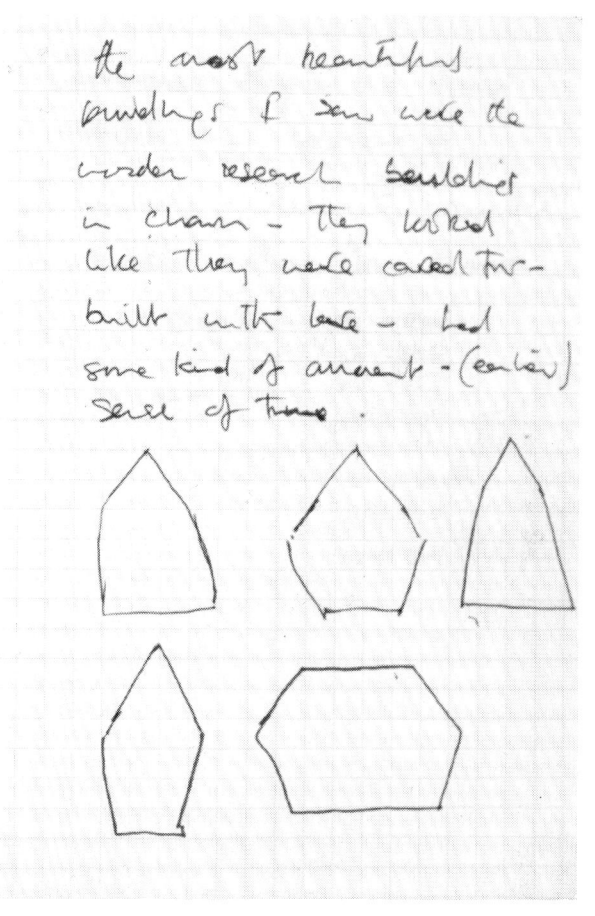

[Handwritten notes on graph paper, partially legible:]

*the most beautiful
buildings I saw were the
wooden research buildings
in Cham— they looked
like they were carved for—
built with love— had
some kind of ancient — (earlier)
sense of time.*

[Sketches of six building silhouettes/rooflines]

Valera, the artist we are visiting, is a landscape painter. The smallish naturalistic paintings are hung on his studio wall against a background of faded floral wallpaper. Valera has made many expeditions into the mountains of eastern Siberia and he painted *plein air* what he sees. He also makes videos of the expeditions, straight forward records, no frills and without art. So we spend the evening watching videos of expeditions to Kamchatka, Jack London Lake and the area around Yakutsk. It is a strange sensation to be watching videos in Magadan, and I ask myself the question, what difference does it make to watch them here or at home. Then realise as the night goes on that here things are gradually going Magadan.

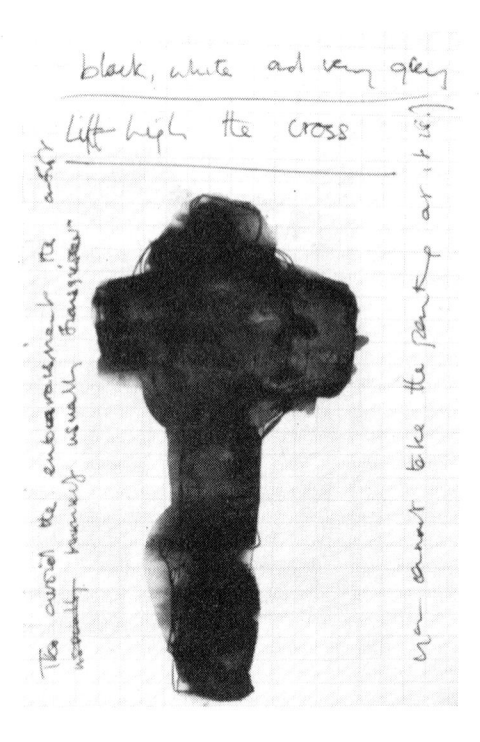

The preceding story-essay and illustrations are part of a co-operative project organised by the Pori Art Museum in Helsinki, which brought together a group of 'cultural scientists' with three artists: Ian McKeever, a Finn, Lauri Antilla, and a Norwegian, Marianne Heske, to explore our culturally constructed concepts of nature.

In August 1992, the group travelled to the biological research station of Chaun, on the Chukotkan peninsula in Eastern Siberia, near the polar circle. This exploration of wilderness landscape was in one sense an attempt to enter a place untouched by man; yet at the same time, their imaginative and physical explorations entailed a recognition that the place was already imprinted with history, with myriad migratory routes crossing between the two continents and continents and ancient traditional cultures; and more recently, tragically, with the Gulag.

The project's exhibition *Ikijaa/Permafrost* took place at Pori Art Museum in 1992. An extensive catalogue of the entire project was published by the Museum in 1995.

Ian McKeever was born in Withersea, East Yorkshire, in 1946. He is an autodidact who now lives and works as an artist in Hartgrove, Dorset. His most recent exhibition was at the Angel Row Gallery, Nottingham.

Dwelling in the North[1]

Tony McManus

> ... when any given environment or function, however apparently 'productive', is really fraught with disastrous influence to the organism, its modification must be attempted or, failing that, its abandonment faced ...
>
> PATRICK GEDDES

KENNETH WHITE'S WAY BOOKS are prose-narratives in which the protagonist starts in the urban context of the twentieth century West and moves from there into the landscape, through moments of clarification and illumination which seem to expand the protagonist's presence in the world, way beyond the confines of urban life and the time frame of the present. He becomes, in a peculiar unification of the material with the abstract, simply 'human presence in the earth', *residence*[2] in the world. In an early neologism, White latinised and grecified this phrase to 'bio-cosmo-poetics', which eventually evolved into *geopoetics: a poetic dwelling in the earth.*

The Wild Swans[3] was originally published in Paris as *Les Cygnes Sauvages*, a title which contains a key French pun on the word 'signes', or 'signs'. The narrative is ostensibly a journal of Japan, from the urban hell of Tokyo northwards to the 'skate-shaped island' of Hokkaido – 'north sea way', where the narrator encounters first hand a long held interest in the remnants of the Ainu culture. He has a dream in mind: '... to see the wild swans coming swooping and whooping from Siberia to winter in Japan's northern lakes ...'.

White's journey follows the footsteps of the great Japanese writer, Matsuo Basho (1644-1694), who made a similar journey with a dream in mind of '... the full moon rising over the islands of Matsushima ...'. On another level, Basho's 'journey into the deep north'[4] is the journey towards a more substantial, clarified poetics – it had fallen to him to become the re-maker of Japanese literature, which he believed to be moribund, mere style and mannerism weighed down by metaphor and word-play, lacking substance.

When Matsuo Basho reached the barrier gate at Shirakawa, where the tradition was that if you entered into the northern territories there, you wrote a poem to mark your passage, he couldn't write anything. The master of literature, lost for words! But he wasn't concerned. He absorbed the scene, *lived* the poem of the earth at Shirakawa:

> After many days of solitary wandering, I came at last to the barrier-gate of Shirakawa, which marks the entrance to the northern regions. Here for the first time, my mind was able to gain a certain balance and composure, no longer a victim to pestering anxiety, so it was with a mild sense of detachment that I thought about the ancient traveller who had passed through this gate with a burning desire to write home. This gate was counted among the three largest checking stations, and many poets had passed through it, each leaving a poem of his own making. I myself walked between trees laden with thick foliage, with the distant sound of autumn wind in my ears and the vision of autumn tints before my eyes. There were hundreds and thousands of pure white blossoms of *unohana* in full bloom on either side of the road, in addition to the equally white blossoms of brambles, so that the ground, at a glance, seemed to be covered in early snow.[5]

And when a fellow-writer asks him what he had written at the gate it makes him think about the nature of poetry itself. 'What *is* poetry?', he asks himself. 'What was my first encounter with poetry?' And, at last, in answer to this question, he writes his Shirakawa barrier-gate poem:

> The first poetic venture
> I came across –
> the rice-planting songs of the far North.

In this, Basho is echoing the words of the great Chinese poet Wang Wei (706-761), who lived one thousand years previously. Wang Wei was asked by the Magistrate Chang what was the 'ultimate truth'. In other words politics asks poetry for the 'truth'. Wang Wei's answer is a beautiful unification of *residence* and *errance*:

> REPLY TO THE MAGISTRATE CHANG
>
> Evening, only quietness pleases me
> the world has gone from my mind
> now it is mine the future means nothing to me
> I know how, like the ancient forest, to return.

> The wind in the pine-trees unknots my belt
> my lute chimes in the twilight of the peaks
> Ultimate truth? –
> the song of the fisherman out among the reeds.[6]

'Ultimate truth? – the song of the fisherman out among the reeds' – again this is an image of the human presence in the world, hearing, and giving voice to the world, like Basho's 'rice planting songs of the far North'.

Moving across our northern territory to these isles, about 150 years prior to Wang Wei, we hear another echo of the same theme in Taliesin:

> I am the wind that blows over the sea
> I am a wave of the sea
> I am the roaring of the sea
> I am a bird on the seacliff ...

For Taliesin there is no personal identity – his voice expresses an *enlargening* of identity into the landscape. This introduces another northern strain into our hyperborean song – the pictish-druidic – a strain which informed Celtic Christianity and made it so distinctive as to arouse the enduring scorn of establishment western culture, as a threat to be mocked with the accusation of 'stupidity'.

Further west on our journey round the north, and further back in time to the archaic, we find another echo of this strain in the great poem of the Iglulik woman shaman, Uvavnuk:

> The great sea moves me
> it sets me adrift
> it moves me as a reed on the river
> sky and storms
> move the spirit in me
> and I am carried away
> trembling with joy

She enters the landscape-seascape of her earth – becomes it as it becomes her – in an erotic image underlying all these voices, it makes her 'tremble with joy'. In Basho, Wang Wei, Taliesin and Uvavnuk, human expression is expression of the landscape, the voice of the earth. They have nothing whatever to do with notions such as 'self' or 'ego', there is no 'opinion', no concern with 'society', no 'angst'. For them, poetics is the realm of more refined and more expansive perception, poetry the voicing of presence in and of the world.

In this sense, human expression is akin to other expressions of the landscape – tree, water, stone, bird, beast – *not* artificial. It is, or can be, *natural* as a bird's cry, and genuine poetics seeks the context in which human expression can once more be 'natural' in this way. Basho's, Uvavnuk's, Taliesin's and Wang Wei's are voices of an utterly different cultural outlook to the graeco-roman-judaic-mediterranean culture which grounds what we understand as the 'west'. Basho sees the human *in* the world; the west sees the human *and* the world. This latter outlook is most neatly outlined in Descartes' formula – 'res cogitans, res extensa' – an outlook which is reflected in the Subject-Verb-Object pattern of the very languages we speak.

The great movement of American poetry in the 20th century was given impetus by Ernest Fenollosa and Ezra Pound's exploration of the cultures and languages of China and Japan. For Fenollosa, the syntax of western languages was inimical to the world they sought to describe:

> A true noun, an isolated thing, does not exist in nature. Things are only the terminal points, or, rather, the meeting points, of actions, cross-sections cut through actions, snapshots. Neither can a pure verb, an abstract notion, be possible in nature. The eye sees noun and verb as one: things in motion, motion in things.[7]

Which is to reach a similar conclusion from a cultural-linguistic perspective, as the quantum physicists reached from a scientific one. Is what we see made of particles or waves? substance or motion? We cannot tell.[8]

*

It is hardly surprising that people are looking for alternative ways to ground thinking and being. The project of modernity in the west appears to have brought us to a point where we either climax in catastrophe (the Somme, Hiroshima, Chernobyl), or wither away into banality as our abuse of mass communication overloads language and thought. It is no surprise either that Scotland, shackled as it has been to a history obsessed with the social-political, should be slow to accommodate the new thinking required. To listen to the mainstream of Scottish media is to hear a people constantly worrying over national/personal identity, not the enlarged identity of an 'open world'.

One Scot who has engaged with a broadly hyperboreal voice is the Amsterdam-based composer Magnus Robb. He spent some time in Siberia studying, observing and listening to a remarkable bird, the ruby-throated warbler, which imitates others with great accuracy.

Robb wonders if it merely copies what it hears, if sounds are taught to it by its parents, or if they were passed down in the genes, 'recorded in the field' as it were, in previous generations. Robb's composition 'Summoning Dawn: the ruby throat dreaming'[9] is scored for mezzo-soprano and sung in a language created by the composer for the piece. The words are partly invented, partly the names of birds in various historical tongues – Scots 'laverock' (skylark) predominates. Robb's 'language' not only imitates the sound of the ruby-throat, but also attempts to recreate its syntactical forms.

This Scottish composer's return to the birch forest and 'dwelling in the north' is reminiscent, in its fusion of voice, of the Canadian Glenn Gould's 'The idea of north'. Gould was a pianist so highly regarded as to be deemed almost perfect, who, having played in the world's concert halls, decided he was going nowhere and shut himself away, residing mainly in a recording studio, where he sought the ultimate renderings – through electronic media – of primarily the music of Bach and Mozart.

As is often the case, the accidental by-product of self-imposed isolation is of enduring interest, and Gould, using the technology at his disposal, put together his *Solitude Trilogy*[10], of which 'The Idea of North' is the first part. This is a composition of five voices speaking, five westerners who for various reasons have lived in the north. We hear their comments and thoughts juxtaposed, sometimes overlapping like instruments of an orchestra. Themes, motifs (motives) come and go through *crescendi* and *diminuendi*, *ralentandi* and *accelerandi*.

The piece is united by two constants – by the sound of a train upon which the speakers are supposed to be travelling northwards from urban Canada to the tundra and taiga of the Northern Teritories; and by a narrator, Wally MacLean, a surveyor who has spent his working life in the north. In the climax, when the other voices have had their say, Wally takes over while the closing of Jean Sibelius' 5th Symphony plays in the background.

MacLean recalls the words of William James, when he said that: 'There is no moral equivalent for war' – James, under the guise of a statement, is setting the question, 'If war is immoral, what might be its moral counterbalance?' 'Very few of us,' says MacLean, 'Can afford to be *for* something.' In other words, peace – unlike war – is not an active process, people do not clash in the same physical way as when they make war, when no account of what can be 'afforded' is taken.

MacLean goes on to say that what humanity has been up against, what its 'common enemy' was, is what has been called 'Mother Nature'. 'Humans ... used to combine against Mother Nature ...'

and there was 'a cleanness' about it now missing in urban culture, which was 'the idea of north'. But 'no longer do humans combine to defy, or to measure, or to read, or to understand, or to live with what they call 'Mother Nature ... our common enemy is no longer *mother* nature but *human* nature. It has crept steadily from the south ... now it is infecting everything with its contagion.' And so Wally MacLean concludes, and Glenn Gould with him, that the 'moral equivalent of war, for us, is going North.'

Now one northener who tackled this philosophical problem around a hundred years ago was Patrick Geddes, that most atypical of thinkers, of whom Lewis Mumford said, 'he was the first person to use the concept of energy in sociology.' Indeed Geddes' whole life can seen as a struggle for an active 'peace', equivalent to the 'moral' state of 'war' which James had noted the absence of:

> Real peace must be an unending fight against disease and slums, ignorance and economic justice, against deforestation and waste of natural resources; such a peace means, both correctly and figuratively, that everyone must care for his garden.

The 'peace' that Geddes describes is redolent with the themes and motifs of our 'poetic dwelling' in the world. It speaks of isolation, of the individual sensing their presence in the world, enlarging identity therein through action. '*Il faut cultiver son jardin – that is the hygiene of peace.*' So there is a responsibility incumbent on the individual.

It speaks, this peace, of an active life, not conditioned to accepting meekly what is doled out but seeking to know one's presence and express that in a language which is alive, not the dead formulae of our conditioning. And returning to the work of Kenneth White, we discover a parallel movement out from the narrow urban context, towards the geopoetic 'open world', away from the 'motorway of western civilization' along the less trodden paths of the intellect, like St Brandan going 'further and further ... into the white unknown,' that opaque adjective implying, maybe, a 'red known'. A movement whose impetus is raw energy, not static idealistic 'faith', nor moral 'principle', nor personal 'identity'; a restoration to our culture of a true 'sense of the world' – and etymology tells us that means 'a sense of the age of man'; the human *in* the earth, an enlargening of identity *beyond* self:

> this pool of water
>   holding rock and sky
>   traversed by the wing flash of birds
>   is more my original face
>   than even the face of Buddha

And he speaks of a need to express this 'sense of world' as a naked presence in the world:

> with one aim in view
> say the world anew
> dawn talk
> grammar of rain, tree, stone
> blood and bone

The 'new' poetics evoked here by White is more exactly an 'original' one (in both senses), as described by Roger Caillois, when he wrote that 'in the beginning, as far as we can judge, poetry, rather than being a sacred language, was a general language.'

Having wandered through these northern spaces, the conclusion I would like to draw is this: that in their evocation of 'human presence in the earth', their advocacy of an energised and illuminated life, their immense perspective of spacetime, 'sense of world' and desire to express that in a language which is 'original', the 'idea of north' which Glenn Gould sought as 'the moral equivalent to war', the active 'peace' which Patrick Geddes espoused, and poetic 'dwelling in' the earth, these things are the same, by other names.

NOTES:

1. This essay is intended as a rough 'topography of Geopoetics', to offer a general survey of the territory Kenneth White has called 'the great work field' of the world, where energy must be devoted in order to achieve clarification, illumination of thought and expression. The epigraph by Patrick Geddes is in *Edinburgh Review 88* (1992) which devoted itself largely to this remarkable figure, an issue which also contains Kenneth White's essay, *Elements of Geopoetics*.
2. Etymologically, 'dwelling' derives from the verb 'to delay', and so implies movement, as the idea of 'dwelling for a moment' conveys; a repose, perhaps a concentration of energies, breaking an ongoing journey. This counterpoint of 'moving' and 'staying' combined in the word 'dwell' is denoted by two key words in White's vocabulary – *errance* and *residence*, which can be thought of as corresponding respectively to the senses of 'dwell on', and 'dwell in'.
3. The English translation is published in *Pilgrim of the Void* (Mainstream, Edinburgh, 1990).
4. Basho, *The Narrow Road to the North*.
5. ibid.
6. G. W. Robinson, in *Wang Wei, Poems* (Penguin Classics, London, 1995) annotates the poem thus: "The Fisherman's Song' is found both in the *Mencius* and in the *Ch'u Tzu* or *Elegies* of Ch'u. The allusion here is presumably to the latter context. Ch'u Yun, a poet of the fourth

century, to whom the poem is traditionally though wrongly attributed, explains to a fisherman why he is wandering in exile by the river: the world is dirty, he alone is clean; everyone is drunk, he alone is sober and so on. The fisherman finally rows away, singing, 'How clean the river water – I will wash my feet.' That is to say, one's behaviour should be adapted to whatever circumstances prevail.' This instruction to attend to the immediate and specific, so that the general malaise is ameliorated, is echoed in Patrick Geddes' desire to see the general principles he believed in, of conservation and caring nurture, made real and particular through slum renovations and small urban gardens, and it implies a geopoetic 'human presence in the world'.

7. *cf.* Heidegger: the western person no longer speaks, but merely opines; a difference illustrated by comparison of the sentence, 'you are that' with 'I see the world'. The former is enlivened by the conflation of subject with object, the identification of the human as *of* and *in* the world. By a different route, quantum physics has come to the same point. Heisenberg indicates that: 'an objective physics ... ie. a sharp division of the world into subject and object, has indeed ceased to be possible'.

8. Modern concern for the kind of language required for the task of describing this 'new' world has shown itself in a number of experimental literary movements which have expressed interest in the way that language works. Gerard Manley Hopkins turned to a poetic theory which sought to capture the actual energies and currents of nature through manipulation of language, and Walt Whitman turned his ear to the long landscapes of native America. Echoing Basho's views on the literature of his timespace, Thoreau writes: 'we are in danger of forgetting the language which all things speak and events speak without metaphor.' In Scotland, C. M. Grieve spawned 'Hugh MacDiarmid', whose output was the volcano of creative energy Grieve needed to mulch a fertile cultural soil where different outlooks and forms of expression could be developed, embracing arcane and archaic vocabularies from earth scientists to sheep-farmers in Perthshire. MacDiarmid hit the geopoetic note in 'On A Raised Beach'.

9. Magnus Robb's 'Summoning Dawn; the ruby throat dreaming' was premiered in Glasgow in 1996, by Linda Hirst.

10. Glenn Gould's *Solitude Trilogy* is available on CD from the Canadian Broadcasting Company.

Tony McManus is the writer/musician/educator, and organiser of the Scottish Centre Geopoetics, who curated a major exhibition of Kenneth White's life and work for the National Library of Scotland. The above essay is culled from an illustrated lecture presented to the Centre in 1997. Further information on the work of the Geopoetic movement is available from the Centre at 23, Buckstone Crescent, Edinburgh.

REVIEWS & SHORTLEET

MARIO RELICH

MICHAEL NEWTON

JOHN BURNSIDE

Voices From War
Ian MacDougall (ed.)
The Mercat Press 1995
ISBN 1873644 450

MARIO RELICH

'I just refused to kill people'. One can imagine Norman MacCaig declaring this in a sardonic, soft-spoken voice. MacCaig was one of twenty-nine men and women interviewed by Ian MacDougall over the past thirty years. *Voices from War* collects these interviews, reproducing them as personal stories in monologue form. The full title, as displayed on the title-page, but not on the cover, adds *And Some Labour Struggles*, the sub-title being 'Personal Recollections of War in our Century by Scottish Men and Women'.

The interviews should be read consecutively, rather than randomly, because the book makes its impact in an incremental, ultimately overwhelming way, a chronicle of memories which scar our dying century. The interviewees come across as fascinating individuals, who reveal their characters with almost novelistic complexity, and often subtlety. It is evident that he is no faceless, condescending sociologist, much more of a listener rather than an interrogator.

Sixteen of the recollections are by those who took part in The Great War and labour struggles of the immediate post-war years. For many of these, the war came as an explosive shock which initiated or intensified a socialist view of how society functioned. As V. G. Kiernan puts it in his foreword, 'In the light of this book's legacies, it would seem that all through at least the first half of this century a smouldering class struggle, a hidden guerilla war, was going on'.

Such hints of 'smouldering class struggle' certainly make the WWI recollections more than just a catalogue of frightful carnage. Peter Corstorphine, a soldier for most of his life, not only took part in the war, but was also sent to suppress the Irish nationalist movement, and the miners' strike of 1921. Another contributor, James Marchbank, is forthright in his plain-speaking about the horrors of war: 'No one has any conception of what really happened unless you were involved in it. My God, you know, it's fearful to come across men in a' sort of situations, standin' up, lyin' down. bent up, sittin' as if nothing had happened – but dead'. Captain John Lauder, son of Harry Lauder, killed in action according to official sources, supposedly calling out the words 'Carry on!' But another story emerges from William Murray, who claims he was very unpopular with his men because 'he jist treated them as outcasts'. Murray recounts how on Christmas Day during the Battle of the Somme the following occurred: 'We were going over the top and he was shot in the back ... I never had any dealings wi' him but the men all knew he had been shot in the back, oh, aye, they all knew. Well, if anybody is shot in the back it was deliberate'.

A classic case of class war merging into something much more brutal and violent was, of course, the Spanish Civil War. John Lochore was a Scottish volunteer with the International Brigades, and his recollections, the only interview focusing on that conflict, exude a sense of adventure which *almost* transcends the ruthless cruelties depicted. Lochore, a Stalinist at the time, rubbed shoulders with the likes of John Cornford, Stephen Spender, and George Orwell. He is rather bitter about the author of *Homage to Catalonia*: 'What has struck me all these years is his outpourings of hatred: where did his love lie? It certainly wasn't for the Spanish people and the country he was so eager to leave.' But one can hardly fault Lochore for his eloquent description of the conflict as 'a war for the survival of decency on earth'.

The interviews with World War II participants display a greater variety of fighting men, with class war much less of a factor, and the 'survival of decency' becoming a more ambivalent concept. Interviewees include the poet J. K. Annand, who served in the Navy, Bill King, a pilot with Bomber Command, and Eddie Matthieson, who endured the Burma campaign. Unlike MacCaig, and he does say he was never close to him, Annand has some positive things to say about the war as he experienced it: 'The war had no influence on my political views but, oh, it gave me a wider experience of different classes of men and of how people reacted to unusual circumstances and so on. It was a good experience.' Bill King, whose reminiscences are immensely informative about strategic bombing, makes no bones about his role: 'As far as we were concerned it wis a job. Ye were telt tae go and bomb. Ye went and bombed. How could it be indiscriminate? We had tae get tae a certain height by a certain time and fly a certain course and drop your bombs on a certain target.' King, incidentally, also wrote poems about the war, like J. K. Annand.

Eddie Matthieson's contribution is the longest, and also by far the most dramatic and vivid. It begins grippingly with his description of a recurring nightmare which he suffered for years after his participation in the Burma campaign. His descriptions of jungle warfare, moreover, capture the tedium, danger, and anxiety generated under tropical heat and humidity most memorably. His observations about Japanese codes of war are familiar enough now, but he does manage a detached, almost objectively anthropological attitude about atrocities from the Japanese side, and yet genuine shame and remorse about similar actions from the British side. After declaring that should similar circumstances arise now, he would opt to be a conscientious objector, he concludes with this stark warning: '... I don't think wars'll ever stop. But the most powerful countries in the world now seem to be gettin' the idea that war is useless and serves no purpose. Unfortunately, smaller countries don't seem to have got the message.'

Seven of the interviewees, including MacCaig himself, were conscientious objectors. William H. Marwick, J. P. M. Millar and Dr Eric F. Dott all suggest that treatment of 'conshies' was considerably harsher in WWI than WWII, though in neither cataclysm were they really officially appreciated. But one of the most significant differences between such dissenters in the earlier and later war was that the largest number of CO's in WWI were not absolute pacifists, but objected to war for markets between capitalist powers. In the case of Dr Dott, he started out as a Christian pacifist, and ended up as a socialist objector. The WWII CO's, on the other hand, appear to have been more thoroughgoing pacifists, some of them having belonged to the pre-war Peace Pledge Union. One of them, Fred Putkeathly, even expresses strong reservations about collaborators were treated. Here is his observation on the treatment of women collaborators in liberated Holland: 'What I didn't like very much was the women with the shaved heads. This is what they did to the woman who had collaborated with the Germans in any way'.

One of the interviewees, Paul Dunand, whom MacDougall had met on a holiday in France, had actually been a French Resistance fighter. He displays no qualms about the treatment of collaborators and, presumably because he was a Gaullist, wastes no opportunity to denigrate Communists in the Resistance. At one point, he comes close to racism in describing the execution of Indian prisoners-of-war who allegedly collaborated with the Germans and committed mass-rape. Summary justice rather than a fair trial was their lot, but Dunand does not seem to notice this: 'They wore a khaki German uniform, but I don't remember seeing them with turbans. Perhaps one or two wore them but in general they had German helmets. Anyway they were brown'. Oddly enough, the only self-confessed fascist of the twenty-nine, Joseph Pia, comes across as much more likeable. He had been interned for being a leading light in an Italian social club in Edinburgh which supported Mussolini. His interview, which concludes the book on a rather lighter note, is rich with descriptions of comic incidents.

Only three women were interviewed by the editor, but they certainly make their mark. Irena Hurny, wife of Mark Hurny, also an interviewee, gives a most grim account of what it was like to be Pole interned by the Soviet Russians. Her account ebbs and flows with the capricious fortunes of war. She, and thousands of other Poles, were freed when in August 1941 Stalin suddenly decided, after Hitler invaded Russia, to treat the Poles as allies rather than enemies. Dorothy Wiltshire gives a brief, but moving account of her CO father during WWI. His stand was based on religious grounds: 'He believed it was wrong to kill and he refused to do so, because in the Bible it is said that you must not kill ... Anyway my father wanted to obey God rather than men. And he stood firm on that.' Grace Kennedy recollects rent strikes in Glasgow, and women's participation in labour struggles, her conclusion a stirring declaration of commitment: 'It was injustice, it was fighting for what is right, and it was injustice that made me a socialist'.

In his introduction, MacDougall apologizes for not having included 'a higher proportion of women's voices', and Kiernan has this to say: 'A shortage of women is to be regretted, though reasons for it are not hard to find. In the bad old days the fact of women being mostly cut off from civic activity and responsibility may well have dulled awareness among them of what was happening in the public arena.' But is Kiernan, however estimable a historian, really right about this?

Lost Kingdoms: Celtic Scotland and the Middle Ages
John Roberts
Edinburgh University Press, 1997
ISBN 0 7486 09105

MICHAEL NEWTON

THE title of this book reveals the author's ambitious goal of reconstructing the early history of Scotland, when the primary cultural fabric of society was Celtic of one sort or another, be it Pictish, Brythonic or Gaelic. The task is very difficult as so few documentary materials remain from this period. The author uses these, as well as place-name evidence and modern scholarship by well-respected historians, to try to synthesise a brief summary of Scotland from its inception to the downfall of Clan Donald at the end of the fifteenth century, especially following the Celtic element which is increasingly eclipsed by the Anglo-Germanic element.

The picture is a very complex one. When Kenneth Mac Alpine, crowned in 843, continued the process already set in motion of consolidating a patchwork of earlier Pictish and Gaelic kingdoms, Vikings were already inflicting considerable losses on the fledgling nation. The Brythonic kingdoms in the south of present-day Scotland had been fighting a losing battle against Anglo-Saxon invaders for more than two centuries. Yet the Mac Alpine kings managed to forge these peoples together and establish the distinguishing features of the nation we still know, although the seats of power and the conflicting interests of those internal provinces never stood still and were a continual source of strife.

Roberts highlights some of the factors which had a role in the formation of allegiances and alliances in the discords of the ages: the ideal of the Scottish nation, marriage ties, fosterage, interpersonal pacts and simple self-interest. Not only did these bonds form between subkingdoms within Scotland but also between subkingdoms and foreign kingdoms, such as Norway, England and France.

Feudalism refashioned the sociopolitical structures of Scotland and had a correspondingly profound impact on the ethno-linguistic makeup. Roberts takes several pages (pp. 36-44) to explain how the settlement of Anglo-Norman lords and their Northern English supporters (and various other foreign hangers-on) into newly created burghs in the 12th century onwards at the invitation of the king was to produce the *Inglis* tongue, later calling itself 'Scots' at the expense of Gaelic. He also explains how place-names reflect the changing status between the native Gaelic speakers and the relatively small band of feudal implementors and how, despite the linguistic shift, Gaelic place names remained or were translated in varying degrees of Anglicisation. The names and origins of these Anglo-Norman families is discussed in several places in the book, sometimes accompanied by a useful genealogical chart to illustrate interdynastic relationships.

The narrative is particularly engaging when recounting the many long episodes of the Wars of Independence. Shifting and intercrossed alliances, as well as old grudges, kept the fires fanned for a long time, and the future fortunes of clans – who emerged at this period –

rose or fell dramatically according to which party they supported in this drama. The Campbells and Clan Donald both greatly benefited from the aid they gave Robert the Bruce, but the conservative nature of Clan Donald in the face of the increasingly Anglicised tendencies of the Scottish élite created the tensions which form the conclusion of the book.

Roberts sticks to a very old fashioned approach to history. His dialectic is primarily concerned with dates, battles and important men. Not only does this make for dry reading, but it overlooks the larger picture of the culture of those Celtic kingdoms of early Scotland which the book is supposed to be about. It certainly is the case that influential leaders have long lasting and far reaching influences on their subjects, but it is also true that the environment in which people live can utterly change them.

The immediate descendants of many of the Anglo-Norman invaders in Ireland were accused of being 'more Irish than the Irish themselves' and the 18th century antiquarian Ramsay of Ochtertyre similarly tells us of similar men in Scotland: 'Notwithstanding their being mostly of Low country extraction, yet in process of time they adopted the language and manners of their Highland vassals, in whom their strength consisted.' Understanding the 'Lost Kingdoms' of Celtic Scotland is not just a matter of discussing the landed gentry and their foreign origins, as is the tendency in this book, but primarily of understanding the ethos of that society into which they were assimilated.

Better acquaintance with Gaelic and Gaelic source material would have greatly improved Roberts' approach. There are poems to the leaders of earldoms and clans which could have livened up the text, exemplified cultural ideals in Heroic Age Scotland and demonstrated the assimilation of leaders of Anglo-Norman origin into this society. He claims (p. 129), with a hint of condescension, that 'Unlike many other Highland clans, the Campbells do not have an ancient pedigree that can be traced back to some distant if half-legendary ancestor, like Niall of the Nine Hostages.' But the Campbells are also called *Duibhnich* in Gaelic from their legendary ancestor Diarmaid O' Duibhne of the Fianna and there are furthermore traditions of King Arthur himself being one of their ancestors! He claims (p. 175) that the actions of Clan Donald were 'not tempered in any way by any sense of allegiance to the Scottish Crown', despite the statement by John MacInnes, one of the foremost Gaelic scholars of this century, that 'there is never, as far as I know, any suggestion in poetry or elsewhere that the Gaels do not owe allegiance to the true line of Malcolm III no matter how much their loyalty might become obscured in the turbulence of history or how much hostility the policies of an individual monarch and the attitudes of the central authorities might provoke.'

There are many simple mistakes that could have been caught by someone familiar with Gaelic. Discussion of personal names is particularly confused. Not only does he not distinguish between English and Gaelic forms – which would have been a useful service – he cites modern English forms of names which did not then exist. There is an especially tortured attempt (p. 167) to explain the origin of the name MacKay which ends with the absurd statement that the name is 'simply a quirk of pronunciation, since *Aedh* came to be spoken as a mere grunt.'

More subtle and problematic is the Whiggish approach to the Celts which, in confident hindsight, portrays them as a primitive culture destined to decline in the face of 'progress'. His depiction of the ancient Celts as barbaric yet artistic obscures more fundamental and mundane aspects of their civilisation and he ascribes Indo-European features to their society which could be found in many contemporary peoples. The back cover claims that after the Wars of Scottish Independence 'Celtic society survived only in the Highlands and Western Isles', contradicting the fact that Gaelic and its associated culture survived in places like Angus, Carrick and Ayrshire into the 18th century. The focus on the defeats at battles and the finale with the downfall of Clan Donald neglects the ongoing influence of the Celtic substratum of Scotland's common inheritance, as recently elucidated by such scholars as John Bannerman and W. D. H. Sellar, and makes its collapse seem like a foregone conclusion.

As if to propagate old stereotypes, the word 'lawless' appears numerous times to describe the Highlands and Highlanders, without an attempt to discuss the 'order in the chaos' of Heroic Age society as explored by scholars in Ireland such as Proinsias MacCana and Nerys Patterson and the inevitable conflicts when it comes into contact with a centralised society with a very different notion of civic authority. On the other hand, Clan Donald had considerable success in maintaining a peaceful social order in its time and he fails to explain why this contrasts so starkly with other periods.

His old fashioned approach is also characterised by a deference to official documentary evidence in preference to information gleaned from Gaelic tradition. These may indeed be treated as two different kinds of information to complement each other and be used with care, but to see dismissive phrases such as 'so called' and 'or so it is claimed' can make the informed reader weary and wary of his biases.

The compilation of historical data to be found in this book is certainly useful as a reference and was doubtless an arduous task. While Roberts has made a valorous attempt to deal with a very difficult subject, I believe that the book does little to progress an understanding of Medieval Celtic Scotland on its own terms. A paradigm shift in perspective and a familiarity with Gaelic and its literature would be necessary for this, and it is surely not too much to expect that professional scholars and historical authors understand the language and culture of the peoples that they research and write about.

Paul Muldoon
Tim Kendall,
Seren Press, £9.95

JOHN BURNSIDE

AT one point in this comprehensive first study, Kendall notes, more or less in passing, Muldoon's 'confession' to wanting to write 'beautifully pellucid, simple lyric poems.' The remark was made in 1994 – at a time when Muldoon's fame had come to rest mainly on his longer, intricate, playful, (or in academic terms, 'ludic') works: poems which 'Ciaran Carson – a friend and admirer – has condemned as Muldoon's "crossword puzzle" strategies'. It would seem that, to a large number of readers, there are really two Paul Muldoons: the game player of *The More a Man Has*, or *Madoc* on the one hand, and the master of such beautiful, pellucid lyrics as *The Fox*, or *Why Brownlee Left*. I have never been altogether convinced by this view: I cannot help but think that Muldoon's particular strength, and the reason why he merits such a prominent place in contemporary poetry, is his ability to combine the playful and the elegiac, the 'crossword puzzle' and the lyric, in ways that no other poet working today could even conceive of doing. It's relatively easy to recognise this strength in poems like *Incantata*, or *Yarrow*, but it's been there all along, and, in its analytical passages, this critical study demonstrates the point admirably.

Kendall has given us a detailed and informed examination of a poet whose work, more than most, lends itself – makes itself available, in fact – to critical analysis. No doubt it is the first of many – as Kendall notes, Muldoon criticism is a rapidly growing field. Most of his observations hit the mark, and only occasionally does the reader stumble over the kind of 'so what' stuff that can make all analysis seem, momentarily, absurd. One good example, with reference to *Madoc*, crops up on page 153: '*The Briefcase* therefore serves as a kind of Chaucerian envoi, but with the opposite wish: Stay little book. The sonnet's rhyme scheme (abcdfgfgedcba) supports this conclusion, by creating what Lucy McDiarmid calls a "watertight poetic form": the central quatrain is surrounded by five symmetrical rhyming lines on each side.' Elsewhere, however, we are in better company: Kendall's ear and eye are pretty trustworthy, and his research appears to have been exhaustive. Reading the book, I was occasionally unsure whether he also subscribed to the 'two Muldoons' theory: he seems to sympathise with the English critic who 'admitted feeling tempted "to throw the whole thing at a computer and say: Here, you do it"' (a reaction which Muldoon gleefully recounts in interview)' and he is quite capable of describing *Madoc* as 'an ugly, opaque, sprawling epic'; elsewhere, however, he applies his mind to the puzzle to great effect, and seems to relish the linguistic games and philosophical pranks as much as the poet himself. Moreover, the final effect of the book is to show, while not asserting in so many words, the full integrity of Muldoon's *oeuvre*.

It's an odd reaction, after all, to even think of throwing a poem like *Madoc* at a computer – a little like using a dictionary or a piece of software to solve a crossword puzzle. Surely the pleasure of any puzzle lies, at least partly, in its difficulty, in the gradual, piecemeal process of revelation and delight. Kendall often bows to the need to give a balanced view: in truth, his heart belongs as much to Muldoon the prankster as to the Muldoon of the pellucid lyric, yet he feels compelled to say, for example, that 'Despite its many strengths, *Meeting the British* is a dissatisfying volume', or that *Yarrow* is 'marginally less exasperating than entertaining': 'this, in spite of fifteen pages of analysis, directly preceding this remark, noting the poem's 'emotional courage', its generosity and its intricacy – fifteen pages in which, he freely admits, it is 'impossible to do justice' to the poem.

Of course, Kendall is merely reflecting a critical position which has become the norm when applied to Muldoon's poetry. Muldoon has the trickster's ability to infuriate, even where he elicits admiration and perhaps this is the key to the problem, (or one of them). I remember, once, hearing him give a talk on his work, at the Hay-on-Wye Festival. As he spoke, he discarded his notes, page by page till, by the time he left the stage, the floor was covered with a litter of A4 paper. It was an ambiguous gesture. On the one hand, as one well known poet observed afterwards, Muldoon appeared to be 'having us on'; on the other, that gesture could just as easily have been read as an indication of modesty, of the poet's unease at being asked to say anything at all about the poetic process, or about his finished work. I read it as a little of both: on the one hand, a challenge to our expectations which was refreshing and revelatory, on the other, a performance, a double take, a slightly (self-)mocking game. Muldoon, of course, is notoriously modest in interviews, as Kendall points out. Nevertheless, in his poetry at least, he has been represented – by some, and not always only by his detractors – as the smart kid in the class, the one who is difficult to like, because he is so damned clever. The smart kid suffers a certain puzzlement at this reaction – after all, he simply delights in his intelligence – and will often deny he's really as smart as people think, ('I'm very bad at any kind of thinking at all – in fact I think I'm probably ... stupid' or 'I've no natural facility for writing anything at all' are some examples from Muldoon interviews quoted here). Yet Kendall wants to like him – or rather, he wants to admit to it. The book concludes with the observation that:

> *Yarrow* must surely stand alongside *The More a Man Has* as Muldoon's masterpiece to date. Marginally less exasperating than entertaining, *Yarrow* challenges our fundamental assumptions of how poetry should be read. That such radically innovative work can already be appreciated, however imperfectly, is sure evidence of Muldoon's poetic genius. As Wordsworth claimed every great and original writer must, Muldoon has himself created the taste by which he is to be relished.

It's somewhat faint praise, considering. But then, it's an odd process in itself, to attempt to isolate two poems, and make them representative of a poet's 'genius': what about that body of marvellous lyrics, which any poet would envy, never mind a poet who has created such radical and vital works 'outside' the lyric? What about *Incantata*, surely one of the most moving poems written in English in our times? Who, having read them, can easily forget the sheer economy of *Trance*, or *Cuba 1962*? The truth is that, although Muldoon is one poet, not two, his genius is manifold. It will be some time, still, before we reach the point for a full assessment of the work, but I feel certain that his reputation will not rest on a couple of 'masterpieces'. Certainly, his expansion of the possibilities of lyric and narrative in poetry will be central to any overall assessment of his position in late twentieth-century literature.

Still, (and setting these quibbles aside), it must be said that Kendall's book is extremely valuable, both in its insights into Muldoon's work, and in its detailed bibliography. The chapters on *Shining Brow*, and *The Prince of the Quotidian / Six Honest Serving Men* are particularly useful, as the first detailed examinations of these works. Elsewhere, we catch the odd glimpse of what might have been an intriguing line of argument: Carroll, for example, is mentioned in passing, in connection with '"Alice A." (as opposed to Alice B. Toklas), who turns out also, in one of her many manifestations, to be Lewis Carroll's mushroom eating Alice in Wonderland.' (An association also taken up by the Jefferson Airplane, in their song *White Rabbit*, with its refrain 'Go ask Alice', who becomes, as the song unfolds, the icon and distanced victim of the hallucinogenic experience.)

Nothing is made, however, of further affinities between Muldoon and Carroll, though both are obsessed with language as play – supreme games players who delight in a twisted, self-referential, simultaneously mad and impeccable logic, compulsive creators of neologisms and portmanteau words, collagists of the absurd and the darkly menacing – just as they both concern themselves with histories of childhood and of their own particular ideas of the 'noble savage'. Other connections and points of reference, (in particular, R.L. Stevenson) are touched upon, though not pursued: occasionally it feels as if Kendall is placing markers, for an eventual return, in what may be a longer and even more comprehensive study and on the evidence of this book, we can only welcome such a work, should he choose to undertake it. In the meantime, students and teachers will find this book invaluable, as a good, well-organised and thoughtful study of a writer whose work can seem daunting to new readers, yet who, nevertheless, amply rewards the effort. Most important of all, the book is a first step towards mending the gap that has appeared, in the eyes of the critics, (though not in the work), between the 'two Muldoons', and showing that this poet's remarkable body of work confirms an extraordinary gift for the seamless melding of the lyrical and the playful.

SHORTLEET

Shortleet will include brief notices of books, magazines and journals of possible interest, some of which may have received scant publicity elsewhere. Not all will be recent publications. Send review copies/advance information to the editors, Edinburgh Review, 22 George Square, Edinburgh EH8 9LF.

Poetry in the British Isles: Non Metropolitan Perspectives
ed. Hans Werner Ludwig and Lothar Fietz
University of Wales Press, Cardiff
ISBN 0 7083 1226 7 320pp. p/b poa

An outstanding collection of essays, introduced by the ever-pugilistic Christopher Harvie, addressing the splintering of British imperialist culture, particularly as evidenced in the poetry of Britain during 'the second half of the twentieth century.' The reader is left in no doubt as to the vitality and proliferation of excellence among the 'peripheral' voices.

The first section – 'Centre and Periphery in Historical Perspective' – assembles three essays, two of which are by the editors, whose place at Tübingen allows them objectivity. The second then focuses the project on 'The Places of Contemporary Poetry' by means of six contributions: two Irish, two Welsh and two Scottish. Ursula Kimpel examines 'Scottish poetry between cultures' in 'Beyond the Caledonian Antisyzygy', while Derick Thomson's fine study of postwar Scottish Gaelic poetry demonstrates how the tradition has redefined itself with new source material. The closing assemblage of work draws the eye in closer through specific studies of five poets: Charles Tomlinson, Tony Harrison, Waldo Williams, Gillian Clarke and George Mackay Brown; the latter, an essay entitled 'The Blinding Breath' by David Anwin, emphasises the importance of the keywords of 'island' and 'community' in the late Orkney poet's worldview.

A serious and successful attempt to assimilate and assess the changing cultural shape of the British island group as its political framework crumbles – proving the poet is often barometer, as well as legislator.

Thirteen Ways of Looking at Tony Conran
ed. Nigel Jenkins
Welsh Union of Writers
ISBN 0 951337 3 5 242pp. p/b £9.00

An interesting collection of pieces creative and analytical in celebration of the 'translator, critic, playwright, and poet of his country's tense', published to coincide with the UK year of Literature and Writing 1995 event. A modernist in the line of Jones, Bunting and MacDiarmid, his contribution to Welsh culture, through urbane and playful Anglo-Welsh writing and work bringing the Welsh 'Taliesin' to an English audience, demands a redrawing of the literary border between two cultures, so mistakably rooted is its source and style. For those unfamiliar with Conran, this volume will provide more than a full introduction.

Negotiating Identity: Rhetoric, Metaphor and Social Drama in Northern Ireland
Smithsonian Institute Press, Washington DC
ISBN 1 56098 520 8 270pp. h/b
£34.75

A fascinating examination of the rural Protestant village of Listymore and the 'urban village' of the Catholic Ardoyne, arguing that Northern Ireland is not composed of two separate cultures, but rather two opposing groups manipulating a shared culture. Metaphors of 'siege' and 'invasion' are unpacked through examination of riots, parades and recreational violence, penetrating even into such 'secret' groups as the Orange 'Royal Black Institution' and the 'Ancient Order of Hibernians.' An illuminating and scholarly study of human contrariness.

The Invisible Reader
Invisible Books, 85 Old Ford Road, London E2 9QD
ISBN 0 0521256 1 7 116pp. p/b
£7.95

Nicely produced, this large-format, illustrated anthology of twenty two writer's work contains mostly poetry, and features, among others, Ian Hamilton Finlay, Peter Plate, Stewart Home, Rob MacKenzie, Harry Gilonis and Anthony Barnett. Images by David Dellafiora and Woodrow Phoenix.

Collected Poems
A.C. Jacobs
Menard/Hearing Eye, London
ISBN 1 874320 10 1 260pp. p/b
£13.99

Arthur Jacobs (1937-1994) came from an orthodox Jewish family resident in the Gorbals before the hammer began to swing, and his early writing, now lost, was in Yiddish. The family moved to London in 1951 and it was in the mid 50s in London that he began to attend Philip Hobsbaum's 'weekly gathering of poets' – the Group, including Lucie-Smith, Peter Porter, Peter Redgrove and George MacBeth, when they were as yet unknown. Jacobs' work as a translator of Hebrew poetry, particularly David Vogel, immediately distinguishes him and although he published little of his own poetry during his lifetime, the contents of this volume should make clear that he is a writer of finely-tuned sensibility, providing an invaluable Jewish perspective on Scottish nationalism and culture. Clearly his upbringing in Glasgow left him scarred to the point of distancing himself from that scene:

> The Glasgow poet's up there
> To show how tough he is.
>
> After all, its what's expected.
>
> Everybody's heard of Rangers and
>   Celtic, the Gorbals
> Gangs and razor fights.
>
> He's out to show them
> What we're made of.
>
> No, son, it'll not do.
> Kick harder,
> Put the boot in more

This publication, including essays by Hobsbaum, John Silkin, John Rety, Anthony Rudolph and Frederick

Grubb, will hopefully begin the reassessment of the work of this previously neglected literary figure, whose 'Judaeo-Scottish matrix' provided the source material for so many of these 'locally grounded, but universally accessible' poems.

Life on a Dead Planet
Frank Kuppner
Polygon, 22 George Square, Edinburgh
ISBN 0 7486 6151 4 198pp. p/b

Kuppner's novels are exactly that in the non-literary sense of the word and this, his first publication following receipt of the now defunct McVitie's Scottish Writer of the Year Award, is no different. Ninety one short prose pieces meld together to form an eyepiece on a nameless euro-Glasgow yawning and struggling for meaningful existence. The narrator, if such there is, button-holes the reader by means of imperative-driven, subjectless phrases, which work to call the reader to the action as a reflexive participant, only to be showered in Kuppner's wry and resigned misanthropy. It is unlikely that Frank Kuppner will ever enjoy the fame of some of his contemporaries in Glasgow, but in time it is not impossible that critics will read in his perfectly honed prose the very essence of the reinvention of Glasgow as a large and cultured European city.

Strange Fish
Magi Gibson & Helen Lamb
Duende, 2 Royal Exchange Court, Glasgow
ISBN 1 900537 03 6 80pp. p/b
£5.99

Part-colour treatment for the work of two emerging Scottish women writers, mixing their poetry with ethereal illustrations by Suzanne Gyseman reminiscent of the Macdonald sisters but with a Chagallian touch of sad whimsy. The poems have a task to get attention among the illustrations, but on examination, both Gibson and Lamb are to be found on the trail of Lochhead and Jamie, handling the everyday in short narrative poems, but yearning for the mystic. Fine production by a new Glasgow publisher.

(g)haun(s)Q
Rodger & Farqurhason
anti-biography, series #4
dualchas, 21 Garturk St. Glasgow
ISBN 0 95214118 6 8 60pp. p/b
poa

The hands of the murderer William Burke tell their horrible story. Rodger and Farqurhason use much typographical play and black 'n' white imagery to carry the guilty to the gallows, as the truth splutters almost incoherently out. Another in dualchas muscly series of experimental writing – see Helsinki further on.

lyrical
rab fulton
HYBRID, 42 Christ Church Place,
Peterlee, Durham
ISBN 1 873412 01 0 19pp. £3.00

Post-Leonard westcoast Scots, swingeing and at times surprisingly sweet. The deceptive title conceals a sequence of trenchant poems based around experiences at Pollock Free State during the RAM74 tree camp. fulton's distinctive mix of sensitivity and bluster shows him as a writer on the radical side, bringing out the voice of new tribalism with a trumpet blast. Dug in.

Gang Doun wi a Sang
Joy Hendry
diehard, 3 Spittal Street, Edinburgh
ISBN 0 946230 25 0 p/b £5.00

Poetic drama borrowing largely from William Soutar's own work to build a picture of this underrated voice from the Scottish Renaissance of the 20s and 30s. The inevitable attention to the tragic circumstances of his illness and early death in Perth doesn't detract from a view of Soutar as a serious thinker and dedicated artist. A fine tribute.

Speaking Likenesses
Maurice Lindsay
Scottish Cultural Press, 130 Leith Walk, Edinburgh
ISBN 1 898218 96 X 55pp. p/b £5.95

A figure in the Scottish landscape for the last-century, Maurice Lindsay has written widely – and continues to do so, it would seem, as he approaches his ninetieth birthday. His style is carefully crafted, poems that look like poems, that rhyme and scan. Though it might be criticised as formally dull, this collection stretches far for subject matter, all brought to rest by Lindsay's distinctively informed and urbane gaze.

The Paix Machine: a novel in Scots
Iain W.D. Forde

Only a few copies of this book have so far produced, at the author's own expense, in which 'Scots was set the challenge of dealing with modern ideas, places technologies and issues'. Such familiar touchstones are welcome where the script draws so on deeply on the Scots Language Society, Concise Scots Dictionary, the preface to Lorimer's New Testament and so on, but surprising the style is fluent and quite easy to read after a time, in a Riddley Walker sort of way. 'Paix', by the way, means 'peace'.

No Hiding Place
Tracey Herd
Bloodaxe Books, P.O Box 1SN, Newcastle NE99 1SN
ISBN 1 85224 381 3 62pp. p/b £6.95

Dundee-based Herd is a storyteller in verse, and oddly the prose-work of Dundonian A.L. Kennedy comes to mind as a useful comparison. Her intimate often first person narratives of dislocated hope and difficult relationships are, like Kennedy's, refreshingly free of nudge-nudge literary allusion and work on a level of real close-in reader identification with persona.

This volume seems to be the work of a natural rapper, able to render the most ordinary moment rhythmic and meaningful, but also of someone more concerned with what language can relate, rather than the concrete, visual dimension of poetry. After reading *No Hiding Place*, one has the lingering feeling that in time Herd the short-story writer will follow from the debut.

Cruising
John Herdman
diehard, 3 Spittal Street, Edinburgh
EH3 9DY
ISBN 0 946230 41 2 68pp. p/b
£4.90

Another in diehard's lengthening list of plays, Cruising is a two-act satirical drama by the author of Imelda and Ghostwriting. Herdman's cast of characters are cast around on the sea journey of Sir Hamish Cadfoot, Lady Cynthia, Wee Davie Cowmeadow and Rev. James Arbuthnot from Edinburgh to Hamburg and Copenhagen aboard MS. Saturn. Class stereotypes are tested and the improbable happens, when Davie lets the secrets of the shipboard games go.

Siusaidh NicNeill
Scottish Contemporary Poets Series
Scottish Cultural Press, 130 Leith Walk, Edinburgh
ISBN 1 898218 55 2 58pp. p/b
£4.95

Nicneill's first collection of poetry draws on her Lewis background and presents a mystical, nature poet in the Gaelic tradition, not so detached as to be above a little satire when necessary, or indeed a serenade to love. One has the feeling that these are poems for the ear rather the eye, that breath will bring them, wriggling alive, right to the hearth of the oral tradition.

Puente #1
Scottish Gaelic Poetry & Hungarian Poetry in translation
ed. Juan Luis Campos, Bill Costley, Tom Hubbard
Contact:
£2.50

The first of a brave new adventure in translation, bridging oceans and language groups, folds out to reveal two posters. Hungarian poetry (Jozsef, Ady, Babits and Aramy translated by Edwin Morgan and Peter Zollman) arranged around as fine a drawing of two neeps as you'll ever see one side – and Gaelic/English dual translations of Frater, Gorman, Neill, macneacail, Mairi Nicdhomaill and Fearghas Mcfhionnlaigh verso. Subscribe now to help secure the bridge.

Thirteen Ways of Looking at the Highlands and more
Colin Will
diehard, 3 Spittal Street, Edinburgh
ISBN 0 946239 48 X 60pp. p/b
£4.90

Striking cover of black stag against red and north sea oil green, with its resonance of 7:84, at first appears to belie the mood within – the stag is black but these are gentler satires and there are other poems besides the title sequence here. But on reading deeper, Will's words have a sharp green edge that slices like a blade of grass, a wisdom about plants and nature from his background in Geological Survey and Botanical Gardens, and a righteous red anger against Mammon.

Standing by Thistles
Anne McLeod
Scottish Contemporary Poets Series
Scottish Cultural Press, 130 Leith Walk, Edinburgh
ISBN 1 898218 66 8 57pp. p/b £4.95

Anne MacLeod's first volume is in part a collection of the occasional, rather than sequenced poems: 'you say I'm not passionate'; generally short, quite personal, if not exactly confessional, in a variety of moods and situations. The effect is that the reader's mind must jump around to follow her wispish thought patterns, but is rewarded for that effort by a warmth and grace within the writing itself.

The second part is the sequence 'Oran Mor': poems of peripheral places, of Lewis and the Tomb of the Eagles. These seem to display a deeper side to the poet's work, where a genuine wonder regarding rootedness and belonging replaces the more flippant earlier mood, as if the open spaces of the edge had inspired her to write beyond the narrower confines of the personal.

Whins
George Gunn
Chapman Publishing, 4 Broughton Place, Edinburgh
ISBN 0 906772 72 9 84pp. p/b £6.95

Gunn's rolling, rambling ampersand-driven style grabs the reader like a benign drunk, saying 'you're listening to me'. Among these intoxicating tales of endless travails through mostly northern Scotland and Ireland, there is no hiding from hard reality in the glow of twilightism – through tradition as a concept is valued, Gunn's Highlands are absolutely now, often harshly lit and peopled by the concerns of the 1990s, pointed up by continual reference to the world beyond and the third parent – media.

In his introduction, Hamish Henderson compares George Gunn to Neruda for his 'passion and inventiveness' and to Dubliner Patrick Kavanagh. These are precedents perhaps, but Gunn is also a Caithness Beat on the road, spontaneous poetry feeling the country's heartbeat and pain – but letting it off with nothing. The blustering, sardonic tone often belies an extreme sensitivity in moments of great lyrical accomplishment. Still a lad, as a maturing writer George Gunn looks set to become one of the key Northern writers of this highly politicised devolution period, as Assynt and Eigg point the way towards a peasant reclamation of land rights.

Nemesis in the Mearns
Love, Laughter and Heartache in the Land of Grassic Gibbon
Clarke Geddes
Scottish Cultural Press
ISBN 1 898218 78 1 225pp. p/b
£9.99

This peculiar book takes the form of a novel involving some of the author's grandparents and some of Grassic Gibbon's *Cloud Howe* characters. Although it has the form of a fairly traditional realist novel, it packs in facts and dates as if they were corroboration of the text's right to exist. But it would be wrong to assume that it is a conscious experiment with intertextuality – it is less that than an unfortunate obsession.

One feels that the author is too apologetic for his own talent, that in doffing the cap to Gibbon, whom he lectures on and clearly adores, he sets his own writing the impossible task of matching up to the master. This is unfortunate because there is evidence that Geddes freed from his heavy debt might have written a fine enough novel in the *roman a clef* mode. As it is, it is too much the child of influence, too retro where Gibbon was writing passionately about recent and contemporary issues.

The Ballad and the Folk
David Buchan
Tuckwell Press, East Linton, Scotland
ISBN 1 898410 67 4 326pp. p/b
£9.99

Paperback reissue of the 1972 classic, with a new introduction by Ian Olson, following Buchan's death at his home in Newfoundland in 1994. Born in 1939 in Aberdeen, where he studied, his masterpiece examines not only the literary style, but the ethnographic context as it varies through the country of the North East – the most fertile ballad area in Britain. A wealth of scholarship in one volume.

The Diary of Patrick Fea of Stove, 1766-96
Transcribed and edited by W.S. Hewison
Tuckwell Press, East Linton, Scotland
ISBN 1 898410 88 7 540pp. h/b
£20

A well-produced and substantial volume containing the daily jottings and accounts of an Orkney farmer over a thirty year period in pre-improvement Sanday. Though the entries are often slight, and there are many references to the weather, the effect of such a long work is that characters and patterns of life begin to emerge and the reader is drawn into a world of great detail and real social interest, through which it is possible to imaginatively reconstruct that past. Includes a fifty page, highly informative introduction by Alexander Fenton.

Water
George Mackay Brown/illus. György Gordon
North & South, 23 Egerton Road, Twickenham TW2 7SL
ISBN 1 870314 27 1 / 28X (signed)
16pp. card cover £3.95/$8.00

A collaboration between the late skald of Hamnavoe and a Hungarian painter combining a sequence of five poems in his best 'kenning' style with a cover colour reproduction and a single pencil sketch. Slight but nicely produced, and a fine reminder of a great talent departed.

The Can-can, Ken?: A Dose o Dorics
John Aberdein
Clocktower Press, Breckan, Stenness, Orkney
ISBN 1 873767 09 9 20pp. pamphlet

Nine shorts in varying degrees of Dee/Donside, Aberdein's biting wit against a bawdy Breughel background serves up some up some of the richest Scots writing around. Pieces vary from eight-line 'Mair Canny Tales' type jokes to memorise for parties, to longer jaunts into Scots jocularity. Another in Duncan McLean's now collectable pamphlet series.

Mindin Rhoda
Shetland Folk Society, Lerwick
ISBN 0 948276 07 55pp. £6.00

Dialect poems and a story, in memory of Rhoda Butler (1929-1994) who turned to poetry as her family grew up and in her later life became Shetland's best loved folk-poet and everybody's granny on Radio Shetland. Here a short biographical essay by Edward Thomason precedes the work of 31 Shetland's poets, mostly all writing in the local tongue which 'Rhoda' herself did so much to revitalise interest in. A tribute to her indeed that so many should now be writing 'i da dialect' at all.

Scottish Skalds and Sagamen
Julian D'Arcy
Tuckwell Press, East Linton, Scotland
ISBN 1 898410 25 9 311pp. p/b £14.99

Julian D'Arcy's study of nine major Scottish writers – Gibbon, Gunn, MacDiarmid, Buchan, Lindsay, Mitchison, Muir, Linklater, Mackay Brown – focuses on the influence of things Norse. Introductory chapters provide an interesting overview of the race debate in Scotland's literature during the first half of the twentieth century, illuminating the disturbing trend to assume a clear Celtic/Norse split – though Gibbon/Mitchell favoured neither, claiming the original Pict for his, and both Gunn and MacDiarmid sought a synthesis.

D'Arcy's thesis that Third Reich appropriation of Nordic imagery brought an abrupt end to most of the Viking sword-waving seems reasonable, but his material is richer with some writers than other – clearly the latter day northern myth-making of Mackay Brown varies in its apoliticality from the work of Linklater, Gunn and MacDiarmid and can't be read in quite the same 'cultural studies' way. But all of the authors emerge from D'Arcy's microscope with motives and prejudice subtly unmasked to some degree, with the cumulative effect that the author has made a distinctive and challenging contribution to study of the period.

Helsinki (not the town)
F.N. O'Gafferty
dualchas anti-biography series #3
ISBN 0 9521418 7 6

A very strange collection of shorts from Hoho Gafferty, which will keep you puzzling for a long time – figuring how many hands he must have, to have so many styles. He certainly has a sense of humour:

> Catarasthenes repels Apateon at the Tron Gate:
> 'Erra gorill'a bawz, free furra poun', great present furra wife!'
> 'Gorilla's bawz yer arse! Them's fuckin' kiwi-fruit!'

The unfamiliarity of strange tropical fruit in northern climes has been a source of humour for a long time – bananas and melons, and so on – now at last the turn of the kiwifruit has come, as a sexual connotation is found.

Selected Poems (1954-1994)
Gennady Aygi, trans. Peter France
Angel Books, 3 Kelross Road, London N5 2QS
ISBN 0-946162-59-X 240pp. p/b £11.95

The first substantial presentation of Aygi's work to an English audience. Gennady Aygi's underground figure is slowly emerging to meet a broader audience. His origins are in the small autonomous republic of Chuvashia, where he is regarded as the national poet, but now in Moscow and ever further afield, he is revered as an experimental poet capable of making, as Edwin Morgan writes, 'the Russian language ... do things it has never done before.'

Aygi's influential and groundbreaking work is showcased in this finely produced bi-lingual edition, with translations by his close friend Peter France, to reveal the poet's development over a forty year period, from his beginnings in the circle of Pasternak to his current standing. We see the evolution of his hallmark style of halting, staccato phrases, free form like social existence to a mystical solitariness in a shifting natural world. An important publication.